T0214189

The Physics of the Manhattan Project

Bruce Cameron Reed

The Physics of the Manhattan Project

Fourth Edition

 Springer

Bruce Cameron Reed
Emeritus, Department of Physics
Alma College
Alma, MI, USA

ISBN 978-3-030-61375-4   ISBN 978-3-030-61373-0   (eBook)
https://doi.org/10.1007/978-3-030-61373-0

This Springer imprint is published by the registered company Springer Nature Switzerland AG
The registered company address is: Gewerbestrasse 11, 6330 Cham, Switzerland

This work is dedicated to my wife Laurie, whose love knows no half-life.

Preface to the Fourth Edition

To readers who have purchased this and possibly earlier editions of this book, I thank you. Your support, suggestions, comments, and criticisms have made each edition better than the preceding one. This will likely be the final edition of *The Physics of the Manhattan Project*, and it incorporates a number of revisions, various new graphs, four new sections, updated data, and clarifications and corrections of a number of points. These revisions reflect my own increased knowledge of the subject matter and a number of helpful suggestions contributed by readers who took time to contact me.

Revisions and new material in this edition include:

- Fundamental data on the *Little Boy* and *Fat Man* bombs are now gathered into a Preamble.
- In Sect. 1.8, the discussion of the number of fission-liberated neutrons above a given energy threshold has been clarified.
- The discussion of the fission barrier in Sect. 1.10 has been revised to bring it closer to Bohr and Wheeler's original analysis. Updated data on fission barriers has been incorporated throughout the book.
- The discussion of tamped criticality in Sect. 2.3 has been significantly improved with the inclusion of several new graphs. An entirely new Sect. 2.4 discusses composite tamped cores, that is, ones comprising shells of two different fissile materials enclosed within a tamper.
- Numerical simulations of the *Little Boy* and *Fat Man* bombs (now Sect. 2.6) incorporate improved physical modeling and updated information on real core and tamper masses.
- Section 2.8 is also new, comprising four subsections on approximate methods of estimating critical masses and yields. These methods are less rigorous than their counterparts in the preceding sections of this chapter but are intended for use in a classroom setting where time restrictions may not permit a fuller development.
- Section 3.3 on plutonium production has been simplified and clarified.
- Section 4.1 on boron contamination in graphite has been significantly revised to correct some errors in previous editions.

- Confusing statements in Sect. 4.3 on predetonation yield have been clarified.
- Section 4.5, also new, describes the physics of the neutron initiators that were developed during the Manhattan Project to trigger *Little Boy* and *Fat Man*.
- Section 5.4, also new, investigates the amusing claim that the energy of fission can make a grain of sand visibly jump.

The fact that you are reading this indicates that you appreciate that the discovery of nuclear energy and its liberation via nuclear weapons was one of the pivotal events of the twentieth century. The strategic and military implications of this development drove much of cold-war geopolitics for the last half of that century, and remain with us today in the form of weapons stockpiles and deployments, proliferation, fissile-material security concerns, test-ban treaties, and concern with the possibility that terrorists or unstable international players might be able to acquire enough fissile material to assemble a crude nuclear weapon. For better or worse, stabilizing or destabilizing, the legacies of the United States Army's "Manhattan Engineer District," Los Alamos, Oak Ridge, Hanford, *Trinity*, *Little Boy*, *Fat Man*, Hiroshima, and Nagasaki will continue to influence events for decades to come, even as the number of deployed nuclear weapons in the world declines.

To sensibly assess information and claims regarding these concerns, one needs some knowledge of the physics that backgrounds nuclear weapons. Should you be merely concerned or downright alarmed if you learn that a potential adversary country is "enriching uranium to 20% ^{235}U" or "developing fuel-rod reprocessing technology"? Why is there is such a thing as a critical mass, and how can one estimate it? How does a nuclear reactor differ from a nuclear weapon? Why can't a nuclear weapon be made with a common metal such as aluminum or iron as its "active ingredient"? How did the properties of various uranium and plutonium isotopes lead to the development of the "gun" and "implosion" weapons used at Hiroshima and Nagasaki? How did the developers of those devices estimate their expected energy yields? How can you arrange to assemble a critical mass in such a way as to avoid blowing yourself up beforehand? This book is an effort to address such questions at about the level of a junior-year undergraduate physics student.

This work grew out of three courses that I taught at Alma College, supplemented with information drawn from a number of my own and other published research articles. One of the courses was a conventional undergraduate sophomore-level "modern physics" class for physics majors which contained a unit on nuclear physics. Another was an algebra-level general-education class on the history of the making of nuclear weapons in World War II, and the third was a junior-level topics class for physics majors that used much of the present volume as its text. What originally motivated this book was that there seemed to be no one source available for a reader with a college-level background in physics and mathematics who desired to learn something of the technical aspects of the Manhattan Project in more detail than is typically presented in conventional texts or popular histories. As my own knowledge of these issues grew, I began assembling a collection of derivations and results to share with my students, and which evolved into the present volume.

I hope that readers will discover, as I did, that studying nuclear weapons is not only fascinating in its own right, but is an excellent vehicle for reinforcing understanding of many foundational areas of physics such as energy, electromagnetism, dynamics, statistical mechanics, modern physics, and, of course, nuclear physics—as well as their attendant mathematical techniques.

This book is neither a conventional text nor a work of history. I assume that readers are already familiar with the basic history of some of the physics that led to the Manhattan Project and how the Project itself was organized (Fig. 1). Excellent background sources are Richard Rhodes' *The Making of the Atomic Bomb* (1986) and F. G. Gosling's *The Manhattan Project: Making the Atomic Bomb* (1999); I also humbly recommend my own *The History and Science of the Manhattan Project* (2019), which fills in much more of the background scientific discoveries and history of Project administration than are covered in the present book. While I include some background material in this book for sake of a reasonably self-contained treatment, I assume that within the area of nuclear physics, readers will be familiar with concepts such as reactions, alpha and beta decay, *Q*-values, fission, isotopes, binding energy, the semi-empirical mass formula, cross sections, and the concept of the "Coulomb barrier." Familiarity with multivariable calculus

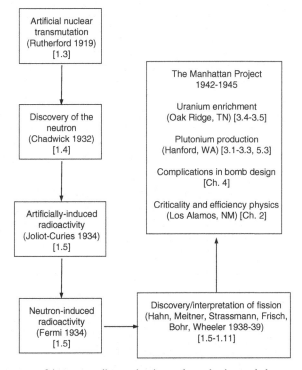

Fig. 1 Concept map of important discoveries in nuclear physics and the organization of the Manhattan Project. Numbers in square brackets indicate sections in this book where corresponding topics are discussed

and simple differential equations is also assumed. In reflection of my own interests (and understanding), the treatment here is restricted to World War II-era fission bombs. As I am neither a professional nuclear physicist nor a weapons designer, readers seeking information on postwar advances in bomb and reactor design and issues such as isotope separation techniques will have to look elsewhere; a good source is Garwin and Charpak (2001). Similarly, this book does not treat the *effects* of nuclear weapons, for which authoritative analyses are available (Glasstone and Dolan 1977). For readers seeking more extensive references, an annotated bibliography appears in Appendix J.

This book comprises 34 sections within five main chapters; Chap. 6 comprises several appendices. Chapter 1 examines some of the history of the discovery of the remarkable amounts of energy released in nuclear reactions, the discovery of the neutron, and characteristics of the fission process. Chapter 2 details how one can estimate the critical mass of fissile material necessary for both untamped and tamped fission weapons, and the explosive yield one might expect of a weapon. Aspects of producing fissile material by separating uranium isotopes and synthesizing plutonium are taken up in Chap. 3. Chapter 4 examines some complicating factors that weapons engineers need to be aware of. Some miscellaneous calculations comprise Chap. 5. Useful data are summarized in Appendices A and B, and a number of background derivations are gathered in Appendices C–G. For readers wishing to try their own hand at calculations, Appendix H offers a number of exercises, with answers provided. A number of symbols are used in this text to designate different quantities in different places, and a Glossary of the most important ones appears in Appendix I. A bibliography for further reading is offered in Appendix J, and some useful constants and conversion factors appear in Appendix K. The order of the main chapters, and particularly of individual sections within them, proceeds in such a way that understanding of later ones often depends on knowledge of earlier ones: Physics is a vertically-integrated discipline.

I have developed spreadsheets for carrying out a number of the calculations described in this book, and which are intended to be used in working some of the exercises in Appendix H. I used to maintain a website from which these were obtainable, but that was becoming expensive. Even institutional links can be volatile, so I ask readers who are interested in obtaining the spreadsheets to contact me directly at reed@alma.edu: I intend to be around for a while. When spreadsheets are discussed in the text, they are referred to in **bold** type.

It should be emphasized that there is no material in the present work that cannot be gleaned from publicly-available texts, journals, and websites: I have no access to classified material.

For many years now, I have benefitted from spoken and electronic conversations, correspondence, suggestions, willingness to read and comment on draft material, and general encouragement from John Abelson, Joseph-James Ahern, John Altholz, Dana Aspinall, Albert Bartlett, Jeremy Bernstein, Alan Carr, David Cassidy, John Coster-Mullen, Steve Croft, Peter Dawson, Gene Deci, Eric Erpelding, Patricia Ezzell, Charles Ferguson, Henry Frisch, Ed Gerjuoy, Chris Gould, Dick Groves, Robert Hayward, Dave Hafemeister, Art Hobson, William

Lanouette, Irving Lerch, Harry Lustig, Mike Magras, Jeffrey Marque, Albert Menard, Tony Murphy, Robert S. Norris, Mike Pearson, Peter Pesic, Sean Prunty, Klaus Rohe, Bob Sadlowe, Tom Semkow, Frank Settle, Ruth Sime, George Smith, D. Ray Smith, Roger Stuewer, Arthur Tassel, Linda Thomas, Michael Traynor, Mark Walker, Alex Wellerstein, Bill Wilcox, John Yates, and Pete Zimmerman. A few of these individuals are, sadly, no longer with us. If I have forgotten anybody, know that you are in this list in spirit. Alma College interlibrary loan specialists Susan Cross and Angie Kelleher have never failed to dig up any obscure document which I have requested; they are true professionals. I am particularly grateful to Steve Croft and Klaus Rohe for a number of helpful comments and suggestions, and to John Coster-Mullen for permission to reproduce his beautiful cross-sectional diagrams of *Little Boy* and *Fat Man* which appear in the Preamble.

Students in my advanced-level topics class—Charles Cook, Reid Cuddy, David Jack, and Adam Sypniewski—served as guinea pigs for much of the material in this book, and took pleasure in pointing out a number of confusing statements, Of course, I claim exclusive ownership of any errors that remain. I am also grateful to Alma College for having awarded me a number of Faculty Small Grants over the years in support of projects and presentations involved in the development of this work. Angela Lahee and her colleagues at Springer deserve a big nod of thanks for continuing to support this project.

Most of all I thank Laurie, who continues to bear with my Manhattan Project obsession.

Suggestions for corrections and additional material will be gratefully received. I can be reached at reed@alma.edu.

Halifax, Nova Scotia   Bruce Cameron Reed
July 2020

References

Garwin, R. L., Charpak, G.: Megawatts and Megatons: A Turning Point in the Nuclear Age? Alfred A. Knopf, New York (2001)

Glasstone, S., Dolan, P. J.: The Effects of Nuclear Weapons, 3rd edition. United States Department of Defense and Energy Research and Development Administration, Washington (1977)

Gosling, F. G. The Manhattan Project: Making the Atomic Bomb. United States Department of Energy, Washington (1999). Freely available online at https://www.osti.gov/biblio/1330716-manhattan-project-making-atomic-bomb-edition

Reed, B. C.: The History and Science of the Manhattan Project, 2nd ed. Springer, Berlin (2019)

Rhodes, R.: The Making of the Atomic Bomb. Simon and Schuster, New York (1986)

Preamble

Fundamental Data on *Little Boy* and *Fat Man*

Throughout this book, reference will frequently be made to the ^{235}U-fueled Hiroshima "gun-type" *Little Boy* bomb and the ^{239}Pu-fueled Nagasaki "implosion-type" *Fat Man* bomb. For easy reference, fundamental data on these devices is summarized in this Preamble. Technical terms used here are developed in more detail in later chapters.

The diagrams of the bombs appearing in Figs. P.1 and P.5 were prepared by and are used with the kind permission of John Coster-Mullen. John's fantastic book on the designs of the bombs and the bombing missions should be on the shelf of any serious student of the Manhattan Project [Coster-Mullen (2016)].

P.1 Little Boy

The core of this bomb comprised two pieces of ^{235}U that were assembled into a supercritical whole by propelling a hollow cylindrical "projectile" piece of ^{235}U initially located in the tail of the bomb forward to seat with a solid cylindrical "target" piece of ^{235}U located in the nose. The target and projectile pieces were contained within what was essentially the barrel of an artillery cannon, and upon assembly were surrounded by a tungsten-carbide (steel; WC) tamper. *Little Boy* is drawn to scale in Fig. P.1, and the projectile seating arrangement is sketched in Fig. P.2. In World War II, the highest velocity that could be achieved for an artillery shell was about 1000 m s^{-1}. The target and projectile pieces were on the order of 20 cm in length (more precise dimensions are given below), which means that the time required for the projectile to become fully mated with the target piece from the time that the leading edge of the projectile met the target piece was ~ 200 ms. This will prove to be an important number.

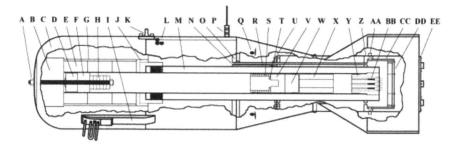

Fig. P.1 Cross-sectional drawing of Y-1852 *Little Boy* showing major components. Not shown are radar units, clock box with pullout wires, barometric switches and tubing, batteries, and electrical wiring. Numbers in parentheses indicate quantity of identical components. Drawing is to scale. Copyright by and used with kind permission of John Coster-Mullen. **A** Front nose elastic locknut attached to 1-inch diameter Cd-plated draw bolt. **B** 15.125-inch diameter forged steel nose nut. **C** 28-inch diameter forged steel target case. **D** Impact-absorbing anvil with shim. **E** 13-inch diameter 3-piece WC tamper liner assembly with 6.5-inch bore. **F** 6.5-inch diameter WC tamper insert base. **G** 14-inch diameter K-46 steel WC tamper liner sleeve. **H** 4-inch diameter U-235 target insert discs (6). **I** Yagi antenna assemblies (4). **J** Target-case to gun-tube adapter with 4 vent slots and 6.5-inch hole. **K** Lift lug. **L** Safing/arming plugs (3). **M** 6.5-inch bore gun. **N** 0.75-inch diameter armored tubes containing priming wiring (3). **O** 27.25-inch diameter bulkhead plate. **P** Electrical plugs (3). **Q** Barometric ports (8). **R** 1-inch diameter rear alignment rods (3). **S** 6.25-inch diameter U-235 projectile rings (9). **T** Polonium-beryllium initiators (4). **U** Tail tube forward plate. **V** Projectile WC filler plug. **W** Projectile steel back. **X** 2-pound Cordite powder bags (4). **Y** Gun breech with removable inner breech plug and stationary outer bushing. **Z** Tail tube aft plate. **AA** 2.25-inch long 5/8-18 socket-head tail tube bolts (4). **BB** Mark-15 Mod 1 electric gun primers with AN-3102-20AN receptacles (3). **CC** 15-inch diameter armored inner tail tube. **DD** Inner armor plate bolted to 15-inch diameter armored tube. **EE** Rear plate with smoke puff tubes bolted to 17-inch diameter tail tube

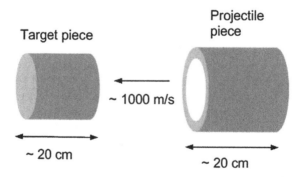

Fig. P.2 Schematic illustration of projectile/target assembly process

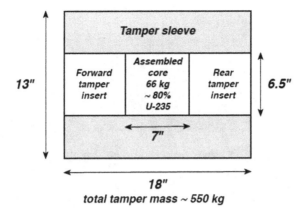

Fig. P.3 Side-view cross-sectional sketch of dimensions of the assembled *Little Boy* core. Not to scale

Table P.1 Data on fissile and tamper materials. A is atomic weight and ρ is density

Material	A gr mol^{-1}	ρ gr cm^{-3}
^{235}U	235.04	18.71
^{239}Pu	239.05	15.6
Aluminum	26.982	2.699
Beryllium oxide	25.01	3.02
Depleted U (DU; ^{238}U)	238.05	18.95
Lead	207.2	11.35
Tungsten carbide (WC)	195.85	15.63

From figures given by Cotser-Mullen, the assembled core was 6.25 inches in diameter, but the tamper bore within which it resided was 6.5 inches in diameter and 7 inches long. The enclosing tamper was 18 inches long by 13 inches in diameter. These dimensions are sketched in Fig. P.3, which is not to scale. With these dimensions, the assembled core had a volume of 214.76 in^3 = 3519.24 cm^3, and the tamper a volume of 2156.9 in^3 = 35345.3 cm^3. With respective densities of 18.71 and 15.63 gr cm^{-3}, their masses were M_{core} = 65.8 kg and M_{tamp} = 552.5 kg. However, it is estimated that the core of this device comprised only ~80% ^{235}U, with the balance being ^{238}U, so in many calculations I will use an operative mass of 0.8(65.8 kg) ~ 52.6 kg, rounded to 53 kg. Data on fissile and tamper materials are gathered in Table P.1. Figure P.4 shows some *Little Boy* test bombs.

Fig. P.4 *Little Boy* test units. *Little Boy* was 126 inches long, 28 inches in diameter, and weighed 9,700 pounds when fully assembled. Photo courtesy Alan Carr, Los Alamos National Laboratory

P.2 Fat Man

In contrast to *Little Boy*, *Fat Man* was a spherical-geometry implosion weapon wherein a subcritical core of ^{239}Pu was squeezed to critical density by explosives configured to crush tamper shells which surrounded the core. Figure P.5 shows a cross-sectional diagram of *Fat Man*. At the very center of the core was an 0.8-inch diameter polonium-beryllium neutron initiator; this is discussed in Sect. 4.5. Working outward from the center, the initiator was surrounded by the ^{239}Pu core proper, of diameter of 3.62 inches. This was surrounded by a shell of depleted uranium (DU; essentially pure ^{238}U) of outer diameter 8.75 inches, which was surrounded by an aluminum shell of inner diameter 9 inches and outside diameter 18.5 inches; the two tamper shells were separated by a thin layer of plastic. Outside the aluminum shell were the explosive charges which created the implosion. Dimensions of these core components are summarized in Table P.2. Photographs of the *Trinity* test implosion bomb and the Nagasaki *Fat Man* combat version can be found in Sect. 4.2.

Table P.2 Data on core/tamper dimensions of *Fat Man*

Component	Inner radius (in)	Outer radius (in)	Mass (kg)
^{239}Pu core	0.4	1.81	6.3
DU tamper shell	1.81	4.375	101
Al tamper shell	4.5	9.25	130

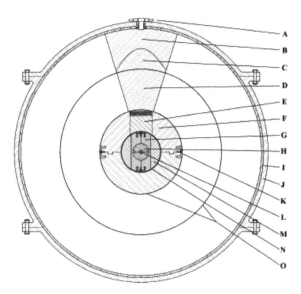

Fig. P.5 Cross-sectional drawing of the Y-1561 *Fat Man* implosion sphere showing major components. Only one set of 32 lenses, inner charges, and detonators is depicted. Numbers in parentheses indicate quantity of identical components. Drawing is to scale. Copyright by and used with kind permission of John Coster-Mullen. **A** 1773 EBW detonator inserted into brass chimney sleeve (32). **B** Comp B component of outer polygonal lens (32). **C** Cone-shaped Baratol component of outer polygonal lens (32). **D** Comp B inner polygonal charge (32). **E** Removable aluminum pusher trap-door plug screwed into upper pusher hemisphere. **F** 18.5-inch diameter aluminum pusher hemispheres (2). **G** 5-inch diameter Tuballoy (U-238) two-piece tamper plug. **H** 3.62-inch diameter Pu-239 hemisphere with 2.75-inch diameter jet ring. **I** 0.5-inch thick cork lining. **J** 7-piece Y-1561 Duralumin sphere. **K** Aluminum cup holding pusher hemispheres together (4). **L** 0.8-inch diameter Polonium-beryllium initiator. **M** 8.75-inch diameter Tuballoy tamper sphere. **N** 9-inch diameter boron plastic shell. **O** Felt padding layer under lenses and inner charges

Contents

About the Author

Bruce Cameron Reed is the Charles A. Dana Professor of Physics (Emeritus) at Alma College in Alma, Michigan. During a 25-year career at Alma, he taught courses ranging from first-year algebra-based mechanics to senior-level quantum mechanics. He earned a Ph.D. in Physics at the University of Waterloo (Canada). His semi-popular text *The History and Science of the Manhattan Project* can be considered a companion volume to the present book. He has also authored a quantum mechanics textbook and has published well over 100 peer-reviewed papers in the areas of astronomy, data analysis, quantum mechanics, nuclear physics, the history of physics, and pedagogical aspects of physics and astronomy. In 2009 he was elected to Fellowship in the American Physical Society (APS) for his work on the history of the Manhattan Project. From 2009 to 2013 he edited the APS's *Physics and Society* newsletter, and served for six years as Secretary-Treasurer of the Society's Forum on the History of Physics. He lives in Nova Scotia with his wife Laurie and a variable number of cats.

Chapter 1
Energy Release in Nuclear Reactions, Neutrons, Fission, and Characteristics of Fission

While this book is not intended to be a history of nuclear physics, it will be helpful to set the stage by reviewing some relevant discoveries. To this end, we first explore the discovery of the enormous energy release characteristic of nuclear reactions, research that goes back to Ernest Rutherford and his collaborators at the opening of the twentieth century; this is covered in Sect. 1.2. Rutherford also achieved, in 1919, the first artificial transmutation of an element (as opposed to this happening naturally, such as in an alpha-decay), an issue we examine in Sect. 1.3. Nuclear reactors and weapons cannot function without neutrons, so we devote Sect. 1.4 to a detailed examination of James Chadwick's 1932 discovery of this fundamental constituent of nature. The neutron had almost been discovered by Irène and Frédéric Joliot-Curie, who unfortunately misinterpreted their own experiments. They did, however, achieve the first instance of artificially induced radioactive decay, a situation we examine in Sect. 1.5; this section also contains a brief summary of events leading to the discovery of fission. In Sects. 1.6–1.11 we examine the process of fission, the release of energy and neutrons during fission, and explore why only certain isotopes of particular heavy elements are suitable for use in fission weapons. Before doing any of these things, however, it is important to understand how physicists notate and calculate the energy liberated in nuclear reactions. This is the topic of Sect. 1.1.

1.1 Notational Conventions for Mass Excess and Q-Values

On many occasions we will need to compute the energy liberated or consumed in a nuclear reaction. Such energies are known as Q-values; this section develops convenient notational and computational conventions for dealing with such calculations.

Any reaction will involve *input* and *output* reactants. The total energy of any particular reactant is the sum of its kinetic energy and its relativistic mass-energy, mc^2. Since total mass-energy must be conserved, we can write

© The Author(s), under exclusive license to Springer Nature Switzerland AG 2021
B. C. Reed, *The Physics of the Manhattan Project*,
https://doi.org/10.1007/978-3-030-61373-0_1

$$\sum KE_{input} + \sum m_{input}c^2 = \sum KE_{output} + \sum m_{output}c^2, \qquad (1.1)$$

where the sums are over the reactants; the masses are the *rest masses* of the reactants. The Q-value of a reaction is defined as the difference between the output and input kinetic energies:

$$Q = \sum KE_{output} - \sum KE_{input} = \left(\sum m_{input} - \sum m_{output}\right)c^2. \qquad (1.2)$$

If $Q > 0$, then the reaction liberates energy, but if $Q < 0$ the reaction demands a *threshold* energy to cause it to happen.

If the masses in (1.2) are in kg and c is in m s^{-1}, Q will emerge in J. However, rest masses are usually tabulated in atomic mass units (abbreviation: amu or just u). If f is the number of kg in one amu, then we can put

$$Q = \left(\sum m_{input}^{(amu)} - \sum m_{output}^{(amu)}\right)fc^2. \qquad (1.3)$$

Q-values are conventionally quoted in MeV. If g is the number of MeV in one Joule, then Q in MeV for masses given in mass units will be given by

$$Q = \left(\sum m_{input}^{(amu)} - \sum m_{output}^{(amu)}\right)(gfc^2). \qquad (1.4)$$

Define $\varepsilon = gfc^2$. With 1 MeV $= 1.602176462 \times 10^{-13}$ J, then $g = 6.24150974 \times 10^{12}$ MeV J^{-1}. Putting in the numbers gives

$$\begin{aligned}
\varepsilon &= gfc^2 \\
&= \left(6.24150974 \times 10^{12}\frac{\text{MeV}}{\text{J}}\right) \times \left(1.66053873 \times 10^{-27}\frac{\text{kg}}{\text{amu}}\right) \\
&\quad \times \left(2.99792458 \times 10^8 \frac{\text{m}}{\text{s}}\right)^2 = 931.494\frac{\text{MeV}}{\text{amu}}.
\end{aligned} \qquad (1.5)$$

More precisely, this number is 931.494013. Thus, we can write (1.4) as

$$Q = \left(\sum m_{input}^{(amu)} - \sum m_{output}^{(amu)}\right)\varepsilon, \qquad (1.6)$$

where $\varepsilon = 931.494$ MeV/amu. Equation (1.6) will give Q-values in MeV when the rest masses are in amu.

Now consider an individual reactant of mass number (= nucleon number) A. The *mass excess* μ of this species is defined as the number of amu that has to be added to A amu (as an integer) to give the actual mass (in amu) of the species:

$$m^{(amu)} = A + \mu. \qquad (1.7)$$

Substituting this into (1.6) gives

$$Q = \left(\sum \left[A_{input} + \mu_{input} \right] - \sum \left[A_{output} + \mu_{output} \right] \right) \varepsilon. \qquad (1.8)$$

Nucleon number is always conserved, $\Sigma A_{input} = \Sigma A_{output}$, which reduces (1.8) to

$$Q = \left(\sum \mu_{input} - \sum \mu_{output} \right) \varepsilon. \qquad (1.9)$$

The product $\mu\varepsilon$ for any reactant is conventionally designated as Δ:

$$Q = \left(\sum \Delta_{input} - \sum \Delta_{output} \right). \qquad (1.10)$$

Δ-values for various nuclides are tabulated in a number of texts and references, usually in units of MeV. The most extensive such listing is published as the *Nuclear Wallet Cards*, and is available from the Brookhaven National Laboratory at www.nndc.bnl.gov; a list of selected values appears in Appendix A. The advantage of quoting mass excesses as Δ-values is that the Q-value of any reaction can be quickly computed via (1.10) without having to worry about factors of c^2 or 931.494. Many examples of Δ-value calculations appear in the following sections.

For a nuclide of given Δ-value, its mass in atomic mass units is given by

$$m^{(amu)} = A + \frac{\Delta}{\varepsilon}. \qquad (1.11)$$

1.2   Rutherford and the Energy Release in Radium Decay

The energy released in nuclear reactions is on the order of a million or more times that typical of chemical reactions. This vast energy was first quantified by Rutherford and Soddy (1903) in a paper titled "Radioactive Change." In that paper, they wrote: "It may therefore be stated that the total energy of radiation during the disintegration of one gram of radium cannot be less than 10^8 g-cal and may be between 10^9 and 10^{10} g-cal... The union of hydrogen and oxygen liberates approximately 4×10^3 g-cal per gram of water produced, and this reaction sets free more energy for a given weight than any other chemical change known. The energy of radioactive change must therefore be at least twenty-thousand times, and may be a million times, as great as the energy of any molecular change."

Let us have a look at the situation using modern numbers. ^{226}Ra has an approximately 1600-year half-life for alpha decay:

$$^{226}_{88}\text{Ra} \rightarrow \, ^{222}_{86}\text{Rn} + \, ^{4}_{2}\text{He}. \qquad (1.12)$$

The Δ-values are, in MeV,

$$
\begin{cases}
\Delta\left({}^{226}_{88}\text{Ra}\right) = 23.669 \\
\Delta\left({}^{222}_{86}\text{Rn}\right) = 16.374 \\
\Delta\left({}^{4}_{2}\text{He}\right) = 2.425.
\end{cases}
\tag{1.13}
$$

These give $Q = 4.87\,\text{MeV}$ in contrast to the *few eV* typically released in chemical reactions.

The notation used here to designate nuclides, ${}^{A}_{Z}X$, is standard in the field of nuclear physics. X is the symbol for the element, Z its atomic number (= number of protons) and A is the nucleon number (= number of neutrons plus number of protons, also known as the atomic weight and as the mass number); this will depend on the particular isotope involved. The number of neutrons N is given by $N = A - Z$.

Rutherford and Soddy expressed their results in gram-calories, which means the number of calories liberated per gram of material. Since $1\,\text{eV} = 1.602 \times 10^{-19}\,\text{J}$, then $4.87\,\text{MeV} = 7.80 \times 10^{-13}\,\text{J}$. One calorie is equivalent to $4.186\,\text{J}$, so the Q-value of this reaction is $1.864 \times 10^{-13}\,\text{cal}$. One mole of ${}^{226}\text{Ra}$ has a mass of 226 grams, so a single radium atom has a mass of 3.75×10^{-22} grams. Hence the energy release per gram is ~ $(1.864 \times 10^{-13})/(3.75 \times 10^{-22})$ ~ 4.97×10^{8} calories, in line with their estimate of 10^{8} to 10^{10}. The modern figure for the heat of formation of water is $3790\,\text{cal}\,\text{gr}^{-1}$; therefore, gram-for-gram, radium decay releases about 131,000 times as much energy as the formation of water from hydrogen and oxygen. In computing the figure of ~ 5×10^{8} calories, we are assuming that the entire gram of radium decays abruptly; in reality, this would take an infinite amount of time and cannot be altered by any human intervention. But the important fact is that individual alpha decays release *millions* of electron-Volts of energy, a fantastic number compared to any chemical reaction.

Another notational convention can be introduced at this point. In this book, reactions will usually be written out in detail as above, but some sources express them in a more compact notation. As an example, in the next section we will encounter a reaction where alpha-particles (helium nuclei) bombard nitrogen nuclei to produce protons and oxygen:

$$
{}^{4}_{2}\text{He} + {}^{14}_{7}\text{N} \rightarrow {}^{1}_{1}\text{H} + {}^{17}_{8}\text{O}.
\tag{1.14}
$$

This can be written more compactly as

$$
{}^{14}_{7}\text{N}\left({}^{4}_{2}\text{He}, {}^{1}_{1}\text{H}\right){}^{17}_{8}\text{O}.
\tag{1.15}
$$

An even more abbreviated notation is ${}^{14}\text{N}(\alpha, p){}^{17}\text{O}$. In this notation, the convention is to have the target nucleus as the first term, the bombarding particle as the first term within the brackets, the lighter product nucleus as the second term within the brackets, and finally the heavier product nucleus outside the right bracket.

1.3 Rutherford's First Artificial Nuclear Transmutation

The discovery that nitrogen could be transformed into oxygen by alpha-particle bombardment marked the first time that a nuclear transmutation was deliberately achieved (Rutherford 1919). This work had its beginnings in experiments conducted by Ernest Marsden in 1915.

In Marsden and Rutherford's experiment, alpha particles emitted by radium bombard nitrogen, producing hydrogen and oxygen via the reaction:

$$\textstyle {}^{4}_{2}\text{He} + {}^{14}_{7}\text{N} \rightarrow {}^{1}_{1}\text{H} + {}^{17}_{8}\text{O}. \tag{1.16}$$

The hydrogen nuclei (protons) were detected via the scintillations they produced upon striking a fluorescent screen. The Δ-values for this reaction are:

$$\begin{cases} \Delta\left({}^{4}_{2}\text{He}\right) = 2.425 \\ \Delta\left({}^{14}_{7}\text{N}\right) = 2.863 \\ \Delta\left({}^{1}_{1}\text{H}\right) = 7.289 \\ \Delta\left({}^{17}_{8}\text{O}\right) = -0.809. \end{cases} \tag{1.17}$$

The Q-value of this reaction is -1.19 MeV. That Q is negative means that this process has a *threshold* of 1.19 MeV, that is, the bombarding alpha must possess at least this much kinetic energy to cause the reaction to happen. This energy is available from the spontaneous decay of radium which gives rise to the alphas; refer to the preceding section where it was shown that decay of ^{226}Ra liberates some 4.87 MeV of energy, more than enough to power the nitrogen-bombardment reaction.

In reality, for reactions with $Q < 0$ the threshold energy is actually greater than $|Q|$ because both energy and momentum have to be conserved; for the above reaction the threshold energy is about 1.53 MeV if the incoming alpha strikes the nitrogen nucleus head-on. The conditions of energy and momentum conservation relevant to head-on "two body" reactions of the general form $A + B \rightarrow C + D$ are detailed in Appendix C. A companion spreadsheet, **TwoBody.xls**, allows a user to input nucleon numbers and Δ-values for all four nuclides, along with an input kinetic energy for reactant A; nucleus B is presumed to be stationary when struck by A. The spreadsheet then computes and displays the Q-value for the reaction, the threshold energy (if appropriate), and the post-reaction kinetic energies and momenta for the products C and D. Of course, most reactions will *not* be head-on, but the point here is to get some sense of the numbers and to be able to make a judgment as to whether or not a transformation is possible in principle. Many nuclear physics textbooks examine the physics of non-head-on collisions, an important aspect of analyzing scattering experiments.

Independent of the Q-value being positive or negative, a related issue in these transmutation reactions that needs to be kept in mind is whether or not the incoming particle has enough kinetic energy to overcome the Coulomb repulsion of the target nucleus and so get close enough to it to allow nuclear forces to come into play; this is examined in Exercise 1.12 in Appendix H.

The physics of two-body collisions is put to considerable use in the following section.

1.4   Discovery of the Neutron

Much of the material in this section is adopted from Reed (2007).

James Chadwick's discovery of the neutron in early 1932 was a critical turning point in the history of nuclear physics. Within two years, Enrico Fermi would generate artificially-induced radioactivity by neutron bombardment, and less than five years after that Otto Hahn and Fritz Strassmann would discover neutron-induced uranium fission. The latter would lead directly to the *Little Boy* uranium-fission bomb, while Fermi's work would lead to reactors to produce plutonium for the *Trinity* and *Fat Man* bombs.

Chadwick's discovery was reported in two papers. The first, titled "Possible Existence of a Neutron," is a brief report dated February 17, 1932, and published in the February 27 edition of *Nature* (Chadwick 1932a). A more extensive follow-up paper dated May 10, 1932, was published in the June 1 edition of the *Proceedings of the Royal Society of London* (Chadwick 1932b). As we work through Chadwick's analysis, these will be referred to as Papers 1 and 2, respectively. The *Nature* paper is reproduced in Andrew Brown's excellent biography of Chadwick; see (Brown 1997). A complete description of the experimental background of the discovery of the neutron would be quite extensive, so only a brief summary of the essentials is given here. A more thorough discussion appears in Chap. 6 of Brown; see also Chap. 6 of Rhodes (1986).

The experiments which lead to the discovery of the neutron were first reported in 1930 by Walther Bothe and his student Herbert Becker, working in Germany. Their research involved studying gamma radiation which is produced when light elements such as magnesium and aluminum are bombarded by energetic alpha-particles emitted in the natural decay of elements such as radium or polonium. In such reactions, the alpha particles often interact with a target nucleus to yield a proton (hydrogen nucleus) and a gamma-ray, both of which can be detected by Geiger counters. A good example of such a reaction is the one used by Chadwick's mentor, Ernest Rutherford, to produce the first artificially-induced nuclear transmutation as was discussed in the preceding section:

$$\ce{^4_2He} + \ce{^{14}_7N} \rightarrow \ce{^1_1H} + \ce{^{17}_8O} + \gamma. \tag{1.18}$$

The mystery began when Bothe and Becker found that boron, lithium, and particularly beryllium gave experimental evidence of gamma emission under alpha bombardment, *but with no accompanying protons being emitted*. The key point here

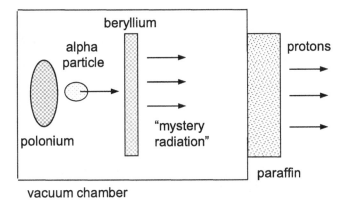

Fig. 1.1 The "beryllium radiation" experiment of Bothe, Becker, the Joliot-Curies, and Chadwick

is that they were certain that some sort of energetic but electrically neutral "penetrating radiation" was being emitted; it could penetrate foils of metal but could not be deflected by a magnetic field as electrically-charged particles would be. Gamma-rays were the only electrically neutral form of penetrating radiation known at the time, so it was natural for them to interpret their results as evidence of gamma-ray emission despite the anomalous lack of protons.

Bothe & Becker's result was picked up by the Paris-based husband-and-wife team of Frédéric Joliot and Irène Curie, hereafter referred to as the Joliot-Curies. In January, 1932, they reported that the presumed gamma-ray "beryllium radiation" was capable of knocking protons out a layer of paraffin wax placed in its path. The situation is sketched in Fig. 1.1, where the supposed gamma-rays are labeled as "mystery radiation."

The energy (hence speed) of the protons could be deduced by determining what thickness of metal foil they could pass through before being stopped, or by measuring how many ion pairs they created in a Geiger counter; such measurements were well-calibrated by Chadwick's time. In comparison to the gargantuan particle accelerators of today, these experiments were literally table-top nuclear physics. In his recreation of the Joliot-Curies' work, Chadwick's experiment involved polonium deposited on a silver disk 1 cm in diameter placed close to a disk of pure beryllium 2 cm in diameter, with both enclosed in a small vessel which could be evacuated; a photograph appears in Brown's biography.

The alpha-producing polonium decay in Fig. 1.1 is

$$\ce{^{210}_{84}Po} \rightarrow \ce{^{206}_{82}Pb} + \ce{^{4}_{2}He}. \tag{1.19}$$

This spontaneous decay liberates $Q = 5.407$ MeV of energy to be shared between the lead and alpha nuclei. The masses of the products involved in such reactions are typically such that their speeds are non-relativistic, a feature we will make considerable use of in our analysis. Even if mass is created or lost in a reaction, momentum must always be conserved. If the polonium nucleus is initially stationary, then the lead and alpha nuclei must recoil in opposite directions. One can easily show from classical momentum conservation that if the total kinetic energy shared by the two product nuclei is Q, then the kinetic energy of the lighter product nucleus must be

$$K_m = \frac{Q}{1 + m/M},$$
(1.20)

where m and M are respectively the masses of the light and heavy product nuclei. Here we have $m/M \sim 4/206$, so the alpha-particle carries off the lion's share of the liberated energy, about 5.3 MeV. The speed of such an alpha-particle is about $0.05c$, justifying the non-relativistic assumption.

We now set up some expressions that will be useful for dissecting Chadwick's analysis.

First, let us assume that Bothe & Becker and the Joliot-Curies were correct in their interpretation that α-bombardment of Be creates gamma-rays. To conserve the number of nucleons involved, they hypothesized that the reaction was

$$_2^4\text{He} + {}_4^9\text{Be} \rightarrow {}_6^{13}\text{C} + \gamma.$$
(1.21)

(Strictly speaking, we are cheating here in writing the reaction in modern notation that presumes knowledge of both neutrons and protons, but this has no effect on the analysis.) From left to right, the Δ-values for this reaction are 2.425, 11.348, and 3.125 MeV, so the Q-value is 10.65 MeV; this energy, when added to the ~ 5.3 MeV kinetic energy of the incoming alpha, means that the γ-ray can have an energy of at most about 16 MeV. However, the energy of the supposed gamma-ray is crucial here, so we do a more careful analysis. In Appendix D, it is shown that if a collision like this happens head-on and if the gamma-ray that is produced travels in the forward direction after the reaction, the energy E_γ of the emergent gamma-ray is given by solving the quadratic equation

$$\alpha E_\gamma^2 + \varepsilon E_\gamma + \delta = 0,$$
(1.22)

where

$$\alpha = \frac{1}{2E_C},$$
(1.23)

$$\varepsilon = 1 - \frac{\sqrt{2E_{He}K_{He}}}{E_C},$$
(1.24)

and

$$\delta = \left(\frac{E_{He}}{E_C}\right) K_{He} - (E_{He} + E_{Be} + K_{He} - E_C),$$
(1.25)

Table 1.1 Δ-Values and Rest Energies for the Joliot-Curie γ-Reaction

Nucleus	A	Δ	E (MeV)
He	4	2.425	3728.40228
Be	9	11.348	8394.79688
C	13	3.125	12112.55116

where E_{He}, E_{Be} and E_C represent the mc^2 rest energies (in MeV) of the alpha-particle, Be nucleus, and carbon nucleus, respectively, and where K_{He} is the kinetic energy of the incoming alpha-particle. These rest energies can be calculated from the corresponding nucleon numbers and Δ-values as $E = \varepsilon A + \Delta$ as in Sect. 1.1.

The relevant numbers appear in Table 1.1.

These numbers give (with $K_{He} = 5.3$ MeV) $\alpha = 4.12795 \times 10^{-5}$ MeV^{-1}, $\varepsilon = 0.983587$, and $\delta = -14.316590$ MeV. Solving the quadratic gives $E_\gamma = 14.55$ MeV; this is a little less than the approximately 16 MeV estimated on the basis of the Q-value alone as the carbon nucleus carries off some momentum. This solution takes the upper sign (+) in the solution of the quadratic; choosing the lower sign leads to a negative value for the kinetic energy of the carbon nucleus, which would be unphysical.

Spreadsheet **TwoBodyGamma.xls** allows a user to investigate head-on reactions of the general form $A + B \rightarrow C + \gamma$. As with **TwoBody.xls**, the user inputs nucleon numbers and Δ-values for nuclides A, B, and C, along with the input kinetic energy for A; B is presumed to be stationary when struck head-on by A. The spreadsheet computes and displays the possible solutions for the energy of the γ-ray and the kinetic energy and momentum of product C.

Returning to the experiment, the 14.6-MeV gamma-rays then strike protons in the paraffin (again assumed head-on), setting them into motion. See Fig. 1.2. Such a collision is a problem in both relativistic and classical dynamics; a γ-ray is relativistic, whereas the protons can be treated classically; this is justified below.

Suppose that the gamma-ray strikes an initially stationary particle of mass m. In what follows, the symbol E_m is used to represent the Einsteinian rest energy mc^2 of the struck particle, while K_m designates its post-collision classical kinetic energy $mv^2/2$;

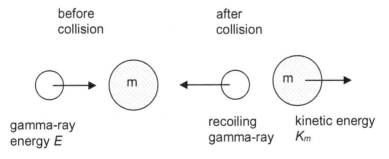

before collision   after collision

gamma-ray energy E recoiling gamma-ray   kinetic energy K_m

Fig. 1.2 A gamma-ray strikes an initially stationary particle of mass m. The latter emerges from the collision with kinetic energy K_m. The gamma-ray is assumed to recoil backwards

E_γ again designates the energy of the gamma-ray before the collision. Maximum possible forward momentum will be imparted to the struck particle if the gamma-ray recoils backwards after the collision, so we assume that this is the case. If the energy of the gamma-ray after the collision is E_γ *, then conservation of mass-energy demands

$$E_\gamma + E_m = E_\gamma^* + E_m + K_m. \tag{1.26}$$

Since we are assuming that the struck particle does not change its identity, the factors of E_m in (1.26) cancel each other. Since the momentum of a photon of energy E is given by E/c, conservation of momentum for this collision can be written as

$$E_\gamma/c = -E_\gamma^*/c + mv, \tag{1.27}$$

where v is the post-collision speed of the struck particle. The negative sign in the right side of (1.27) means that the γ-ray recoils backwards.

It will be useful to also have on hand expressions for the classical momentum and kinetic energy of the struck particle in terms of its rest energy:

$$mv = \sqrt{2mK_m} = \frac{\sqrt{2mc^2 K_m}}{c} = \frac{\sqrt{2E_m K_m}}{c} \tag{1.28}$$

and

$$K_m = \frac{1}{2}mv^2 = \frac{(mc^2)v^2}{2c^2} = \frac{1}{2}E_m\left(\frac{v}{c}\right)^2. \tag{1.29}$$

With (1.28), a factor of c can be cancelled in (1.27); then, on eliminating E_γ * between (1.26) and (1.27), we can solve for E_γ:

$$E_\gamma = \frac{1}{2}\left(K_m + \sqrt{2E_m K_m}\right). \tag{1.30}$$

If the kinetic energy of the struck particle (proton) can be measured, we can use (1.30) to determine what energy the gamma-ray must have had to set it into such motion. On the other hand, if we desire to solve for K_m presuming that E_γ is known, the situation is messier as (1.30) is a quadratic in $\sqrt{K_m}$ that has no neat solution:

$$K_m + \left(\sqrt{2E_m}\right)\sqrt{K_m} - 2E_\gamma = 0. \tag{1.31}$$

We will need an expression for K_m, however, so we must solve the quadratic. Fortunately, we can invoke a simplifying approximation.

The gamma-rays involved here have $E_\gamma \sim 14.6$ MeV, but a proton has a rest energy of about 938 MeV. It is consequently quite reasonable to set $E_m \gg E_\gamma$, in which case this quadratic can be solved approximately, as shown in what follows.

The formal solution of the quadratic is

$$\sqrt{K_m} = \frac{-\sqrt{2E_m} + \sqrt{2E_m + 8E_\gamma}}{2}.$$

Extract a factor of $2E_m$ from under the second radical:

$$\sqrt{K_m} = \frac{-\sqrt{2E_m}\left[1 - \sqrt{1 + 4E_\gamma/E_m}\right]}{2}.$$

Setting $x = 4E_\gamma/E_m$ and invoking the expansion

$$\sqrt{1 + x} \sim 1 + x/2 - x^2/8 + \cdots \quad (x < 1)$$

gives

$$\sqrt{K_m} \sim \frac{1}{2}\left\{-\sqrt{2E_m}\left[1 - \left(1 + 2\frac{E_\gamma}{E_m} - 2\frac{E_\gamma^2}{E_m^2} + \cdots\right)\right]\right\}$$

$$\sim \left\{-\sqrt{2E_m}\left[-\frac{E_\gamma}{E_m} + \frac{E_\gamma^2}{E_m^2} - \cdots\right]\right\}$$

$$\sim \sqrt{2E_m}\left(\frac{E_\gamma}{E_m}\right)\left\{1 - \frac{E_\gamma}{E_m} + \cdots\right\}.$$

Squaring gives

$$K_m \sim 2\left(\frac{E_\gamma^2}{E_m}\right)\left\{1 - 2\frac{E_\gamma}{E_m} + \cdots\right\}. \tag{1.32}$$

This result will prove valuable presently.

Upon reproducing the Joliot-Curie experiments, Chadwick found that protons emerge from the paraffin with speeds of up to about 3.3×10^7 m s^{-1}. This corresponds to $(v/c) = 0.11$, so our assumption that the protons can be treated classically is reasonable. The modern value for the rest mass of a proton is 938.27 MeV. From (1.29), these figures give the kinetic energy of the ejected protons as 5.7 MeV, exactly the value quoted by Chadwick on p. 695 of his Paper 2. Equation (1.30) then tells us that if a proton is to acquire this amount of kinetic energy by being struck by a gamma-ray, then the gamma-ray must have an energy of about 54.4 MeV. But we saw in the argument following (1.25) that a gamma-ray arising from the Joliot-Curies' proposed $\alpha + {}^9\text{Be} \rightarrow {}^{13}\text{C}$ reaction has energy of at most about 14.6 MeV, a factor of nearly four too small! This represents a serious difficulty with the gamma-ray proposal.

Before invoking a reaction mechanism involving a (hypothetical) neutron, Chadwick devised a further test to investigate the remote possibility that 55-MeV gamma-rays might somehow be created in the α-Be collision. In addition to having the "beryllium radiation" strike protons, he also directed it to strike a sample of nitrogen gas. The mass of a nitrogen nucleus is about 14 mass units; at a conversion factor of 931.49 MeV per mass unit, the rest energy of a ^{14}N nucleus is about 13,040 MeV. If such a nucleus is struck by a 54.4-MeV gamma-ray, (1.32) indicates that it should acquire a kinetic energy of about 450 keV. From prior experience, Chadwick knew that when an energetic particle travels through air it produces ions, with about 35 eV required to produce a single ionization (hence yielding "one pair" of ions). A 450 keV nitrogen nucleus should thus generate some 13,000 ion pairs. Upon performing this experiment, however, he found that some 30,000 to 40,000 ion pairs would typically be produced. These figures imply a kinetic energy of $\sim$ 1.1–1.4 MeV for the recoiling nitrogen nuclei, which in turn by (1.30) would require gamma-rays of energy up to $\sim$ 90 MeV, a number completely inconsistent with the $\sim$ 55 MeV indicated by the proton experiment. Indeed, upon letting the supposed gamma-rays strike heavier and heavier target nuclei, Chadwick found that "… if the recoil atoms are to be explained by collision with a quantum [γ-ray], we must assume a larger and larger energy for the quantum as the mass of the struck atom increases." The absurdity of this situation led him to write (Paper 2, p. 697) that "It is evident that we must either relinquish the application of conservation of energy and momentum in these collisions or adopt another hypothesis about the nature of the radiation." To be historically correct, the mass of beryllium atoms had not yet been accurately established in 1932, so Chadwick did not know the $E_\gamma = 14.6$ MeV figure for certain. However, he was able to sensibly estimate it as no more than about 14 MeV unless the beryllium nucleus lost an unexpectedly great amount of mass in the reaction, so, as he remarked in his Paper 2 (p. 693), "… it is difficult to account for the production of a quantum of 50 MeV from the interaction of a beryllium nucleus and an α-particle of kinetic energy of 5 MeV."

The fundamental problem with the gamma-ray hypothesis is that for the amount of energy Q liberated in the α-Be reaction, any resulting gamma-ray will possess much less momentum than a classical particle of the same kinetic energy; the ratio is $p_\gamma / p_m = \sqrt{Q / 2E_m}$, where E_m is again the rest energy of the classical particle. Only an extremely energetic gamma-ray can kick a proton to a kinetic energy of several MeV.

Chadwick's key insight was to realize that if the protons were in reality being struck billiard-ball style by neutral *material* particles of mass equal or closely similar to that of a proton, then the striking energy need only be on the order of the kinetic energy that the protons acquire in the collision.

This is the point at which the neutron makes its debut. Chadwick hypothesized that instead of the Joliot-Curie reaction of (1.21), the α-Be collision leads to the production of carbon and a neutron via the reaction

$$_2^4\text{He} + {}_4^9\text{Be} \rightarrow {}_6^{12}\text{C} + {}_0^1\text{n}. \tag{1.33}$$

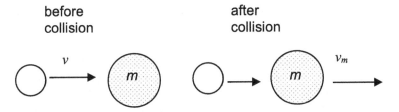

before collision

after collision

v

m

m

v_m

Fig. 1.3 Particle of mass μ strikes a stationary particle of mass m, setting the latter into motion with speed v_m. The sizes of the circles are not meaningful

In this case, a ^{12}C atom is produced instead of the Joliot-Curies' proposed ^{13}C. Since the "beryllium radiation" was known to be electrically neutral, Chadwick could not invoke a *charged* particle such as a proton or electron here. Incidentally, the ^{12}C nucleus will likely remain trapped in the Be target and hence go undetected. If the neutron's true mass and the momentum acquired by the ^{12}C nucleus are accounted for, the kinetic energy of the emergent neutron is about 10.9 MeV. (Chadwick assumed that neutrons were combinations of electrons and protons, and used the mass excess of the proton for his calculations, but this makes little difference to the results.) A subsequent neutron/proton collision will be like a collision between equal-mass billiard balls, so it is entirely plausible that a neutron that begins with about 11 MeV of energy will be sufficiently energetic to accelerate a proton to a kinetic energy of ~ 5.7 MeV even after it (the neutron) batters its way out of the beryllium target and through the window of the vacuum vessel on its way to the paraffin.

As a check on the neutron hypothesis, consider again the nitrogen experiment described above. Instead of a gamma-ray being created in the α-Be collision, presume now a neutral material particle of mass μ (a neutron) is created, which subsequently collides with an initially stationary particle of mass m. This is illustrated in Fig. 1.3.

This collision can be analyzed with the familiar head-on elastic-collision formulae of basic physics; if the neutron has speed v_μ and kinetic energy K_μ, then the post-collision speed and kinetic energy of the struck mass will be

$$v_m = \left(\frac{2\mu}{\mu + m}\right) v_\mu \quad \text{and} \quad K_m = \frac{4\mu m}{(m + \mu)^2} K_\mu. \tag{1.34}$$

Suppose that neutrons emerging from the vacuum vessel do indeed have energies of 5.7 MeV. With neutrons of mass 1 and nitrogen nuclei of mass 14, (1.34) indicates that a nitrogen nucleus should be set into motion with a kinetic energy equal to $56/225 = 0.249$ of that of the incoming neutron, or about 1.4 MeV. This figure is in excellent agreement with the energy indicated by the observed number of ion pairs created by the recoiling nitrogen nuclei!

As Chadwick related in his Paper 2 (p. 698), independent measurements of the recoiling nitrogen nuclei indicated that they acquired speeds of ~ 4.7×10^6 m s^{-1} as a result of being struck by neutrons. Knowing this and the fact that neutron-bombarded protons are set into motion with a speed of about 3.3×10^7 m s^{-1}, he was able to estimate the mass of the neutron by a simple classical argument. If the mass of a proton is 1 unit and that of a nitrogen 14 units, (1.34) indicates that the ratio of the speed of a recoiling proton to that of a recoiling nitrogen would be $(\mu +14)/(\mu +1)$; the measured speeds led him to conclude $\mu \sim 1.15$ with an uncertainty estimated at ~10%. Further experiments with boron targets led Chadwick to report a final estimate of the neutron mass as between 1.005 and 1.008 mass units. The modern figure is 1.00866; the accuracy he achieved with equipment which would now be regarded as primitive is nothing short of awe-inspiring.

In summary, Chadwick's analysis comprised four main points: (1) If the "beryllium radiation" comprises gamma-rays, then they must be of energy ~55 MeV to set protons into motion as observed. (2) Such a high energy is unlikely from an α-Be collision, although not inconceivable if the reaction happens in some unusual way involving considerable mass loss. (3) Letting the same "gamma-rays" strike nitrogen nuclei causes the latter to recoil with energies indicating that the gamma-rays must have energies of ~90 MeV, utterly inconsistent with point (1). (4) If instead the "beryllium radiation" is assumed to be a neutral particle of mass close to that of a proton, consistent results emerge for both the proton and nitrogen recoil energies.

Chadwick was awarded the 1935 Nobel Prize in Physics for his discovery of the neutron. He further speculated in his Paper 2 that neutrons might be complex particles comprising protons and electrons somehow bound together, but this proved not to be the case: Heisenberg's uncertainty principle ruled against the possibility of containing electrons within such a small volume. Subsequent experiments by Chadwick himself showed that the neutron is a fundamental particle in its own right.

1.5   Artificially-Induced Radioactivity and the Path to Fission

Irène and Frédéric Joliot-Curie narrowly missed discovering the neutron in early 1932, but scored a success two years later with their discovery that normally stable nuclei could be induced to become radioactive upon alpha-particle bombardment.

Their discovery reaction involved bombarding aluminum with alphas emitted in the decay of polonium, the same source of alphas used in the neutron-discovery reaction:

$$^{210}_{84}\text{Po} \xrightarrow[138 \text{ days}]{\alpha} {}^{206}_{82}\text{Pb} + {}^{4}_{2}\text{He}. \tag{1.35}$$

The Q-value of this reaction was found in the preceding section to be 5.41 MeV. These alphas then bombard aluminum, chipping off a neutron to leave phosphorus:

$$^{4}_{2}\text{He} + {}^{27}_{13}\text{Al} \rightarrow {}^{1}_{0}\text{n} + {}^{30}_{15}\text{P}. \tag{1.36}$$

The Δ values are respectively 2.425, -17.197, 8.071 and -20.201, which give Q ~ -2.64 MeV. Despite this threshold (negative Q-value), the incoming alpha is more than energetic enough to cause the reaction to proceed. The ^{30}P nucleus subsequently undergoes positron decay to ^{30}Si with a half-life of 2.5 min:

$$^{30}_{15}\text{P} \xrightarrow[2.5\text{ min}]{\beta^+} {}^{30}_{14}\text{Si}. \tag{1.37}$$

It was this positron emission that alerted the Joliot-Curies to the fact that they had induced radioactivity in aluminum. When the bombarded aluminum was dissolved in acid, the small amount of phosphorous created could be separated and chemically identified as such; that the radioactivity carried with the phosphorous and not the aluminum verified their suspicion.

The Joliot-Curies' success stimulated Enrico Fermi to see if he could similarly induce radioactivity by neutron bombardment. He soon succeeded with fluorine:

$$^{1}_{0}\text{n} + {}^{19}_{9}\text{F} \rightarrow {}^{20}_{9}\text{F} \xrightarrow[11.1\text{ sec}]{\beta^-} {}^{20}_{10}\text{Ne}, \tag{1.38}$$

and also with aluminum, discovering a different half-life than had the Joliot-Curies:

$$^{1}_{0}\text{n} + {}^{27}_{13}\text{Al} \rightarrow {}^{1}_{1}\text{H} + {}^{27}_{12}\text{Mg} \xrightarrow[9.5\text{ min}]{\beta^-} {}^{27}_{13}\text{Al}. \tag{1.39}$$

It was not long before Fermi and his collaborators had worked their way through the periodic table to uranium, neutron bombardment of which would lead to the discovery of fission.

Reaction (1.39) is not the only one possible when a neutron strikes aluminum. In such reactions, three different *reaction channels* are typically detected: the neutron may chip off a proton as above, but it may also give rise to an alpha-particle, or be captured by the aluminum nucleus. In all cases the product eventually beta-decays to something stable:

$$^{1}_{0}\text{n} + {}^{27}_{13}\text{Al} \rightarrow \begin{cases} {}^{1}_{1}\text{H} + {}^{27}_{12}\text{Mg} \xrightarrow[9.5\text{ min}]{\beta^-} {}^{27}_{13}\text{Al} \\[4pt] {}^{4}_{2}\text{He} + {}^{24}_{11}\text{Na} \xrightarrow[15\text{ hr}]{\beta^-} {}^{24}_{12}\text{Mg} \\[4pt] {}^{28}_{13}\text{Al} \xrightarrow[2.25\text{ min}]{\beta^-} {}^{28}_{14}\text{Si}. \end{cases} \tag{1.40}$$

If the target is a heavy element such as gold or uranium, the latter channel typically occurs.

The path from the discovery of artificially-induced radioactivity to the discovery of fission was full of near-misses. A brief description of significant developments is given here as a segue into the next six sections, where the physics of fission is covered in more detail. A good qualitative discussion of this material can be found in Chaps. 8

and 9 of Rhodes (1986), and more technical ones in Sime (1996) and in Chaps. 2 and 3 of Reed (2019); a comprehensive technical discussion of developments between the discovery of the neutron and the discovery of fission appears in Amaldi (1984).

As neutron sources, Fermi and his group used small glass vials containing radon gas mixed with powdered beryllium. Radon alpha-decays with a half-life of 3.8 days and is consequently a copious source of alpha particles; these alphas strike beryllium nuclei and produce neutrons of energy ~ 11 MeV as in Chadwick's polonium-source experiment. In early 1934, the Rome group began systematically bombarding various target elements with neutrons. By the spring of 1934 they had come to uranium, for which, at the time, only one isotope was known: ^{238}U. (^{235}U would be discovered by University of Chicago mass spectroscopist Arthur Dempster in 1935.) Upon carrying out the bombardment, they found that β^- activity was induced, with evidence for several half-lives appearing; in particular, they noted one of 13 min. Consequently, they hypothesized that they must be synthesizing a new element, number 93:

$$\,_0^1 n + \,_{92}^{238}U \rightarrow \,_{92}^{239}U \xrightarrow{\beta^-} \,_{93}^{239}X \xrightarrow{\beta^-} ?, \tag{1.41}$$

where X denotes a new, "transuranic" element which might itself undergo a subsequent beta-decay. Chemical testing revealed that their beta-emitters were neither uranium isotopes nor any known element between lead ($Z = 82$) and uranium, a result that strengthened their belief that they were synthesizing new elements. It was in part for this work that Fermi was awarded the 1938 Nobel Prize for Physics.

As it happens, ^{238}U is somewhat fissile when bombarded by very energetic neutrons (see Sect. 1.9). However, the experimental arrangement adopted by Fermi's group precluded their being able to detect the direct ionizations that would be caused by high-energy fission fragments that are so created. In addition to being an alpha-emitter, radon is a fairly prolific gamma-ray emitter, and these gamma-rays would have caused unwanted background signals in ionization-chamber detectors if they were placed near the neutron sources. Consequently, the experimental procedure adopted was to irradiate target samples and then literally run them down a long hallway to a detector far from the neutron source. Since the purpose was to seek delayed effects (induced half-lives are often on the order of minutes), this procedure would presumably not affect the results. However, any fission fragments would have been stopped by then. Fission fragments tend to be neutron rich and suffer a succession of beta-decays, and it must have been beta-decays from such fragments remaining in the bombarded targets that were being detected and attributed to transuranic elements. A common product of fission is barium, and a particular isotope of this element, ^{131}Ba, has a beta-decay half-life of 14.6 min, similar to the 13 min value noted above. Because any reaction that had ever been detected had involved transmutations of elements by at most one or two places in the periodic table, nobody was expecting fission to happen and so never considered that their experimental arrangement might be biasing them against detecting it: Retrospect is always perfect.

In October, 1934, Fermi discovered accidentally that if the bombarding neutrons were caused to be slowed ("moderated") before hitting the target element by having them first pass through water or paraffin, the strength of the induced radioactivities could in some cases be drastically increased. Fermi attributed this to the neutrons having more time in the vicinity of target nuclei, and hence a greater probability of reacting with them. As a result, the Rome group began re-investigating all those elements which they had previously subjected to *fast* (energetic) neutron bombardment. Uranium was one of many elements which proved to yield greater activity upon slow neutron bombardment. Ironically, *slow* neutron bombardment of uranium-238 does create plutonium, which is an excellent material for fueling nuclear weapons; Fermi initially thought this was happening with *fast* neutron bombardment, which tends to lead to fission of this isotope.

The possibility that new elements were being created was treated with some skepticism within the nuclear research community. Among the leaders of that community were Otto Hahn and Lise Meitner at the Kaiser Wilhelm Institute for Chemistry in Berlin, who between them had accumulated years of experience with the chemistry and physics of radioactive elements. In 1935, they and chemist Fritz Strassmann began research to sort out to what elements uranium transmuted under slow neutron bombardment. By 1938, the situation had become extremely muddled: No less than ten distinct half-lives had been identified. To complicate things further, Irène Curie and Paul Savitch, working in Paris, identified an approximately 3.5 hr beta half-life resulting from slow neutron bombardment of uranium, an activity which Hahn and his group had not found. Curie and Savitch suggested that the 3.5 hr decay might be attributed to thorium, element 90. If this were true, it would mean that neutrons slowed to the point of possessing less than one eV of kinetic energy (see Sect. 3.2) were somehow capable of knocking alpha-particles out of uranium nuclei.

Further research by Curie and Savitch showed that the 3.5 h beta-emitter had chemical properties similar to those of element 89, actinium. This observation would eventually be realized as another missed chance in the discovery of fission. To isolate the beta-emitter from the bombarded uranium target, Curie and Savitch used a lanthanum-based chemical analysis. Lanthanum is element 57, which is in the same column of the periodic table as actinium. Chemists were long familiar with the fact that elements in the same column of the table behave similarly as far as their chemical properties are concerned. That the beta emitter "carried" with lanthanum in a chemical separation indicated that it must have chemistry similar to lanthanum, and since the element nearest uranium in the periodic table with such chemistry is actinium, it was assumed that the beta-decayers must be nuclei of that element. The possibility that uranium might in fact be transmuting to lanthanum would have seemed ludicrous, as U and La differ by a factor of nearly two in mass. Curie and Savitch were probably detecting ^{141}La, which is now known to have a half-life of 3.9 h.

Hahn, Meitner, and Strassmann resolved to try to reproduce the French work. Tragically, in July, 1938, Meitner was forced to flee to Holland. Born into a Jewish family in Austria, her Austrian citizenship had protected her against German anti-Semitic laws, but this protection ended with the German annexation of Austria in March, 1938. Hahn and Strassmann carried on with the work and corresponded with her by letter, but her career was essentially destroyed.

By December, 1938, Hahn and Strassmann had refined their chemical techniques and had become convinced that they were detecting barium (element 56) as a result of slow-neutron bombardment of uranium. Barium is adjacent to lanthanum in the periodic table, and is another common product of uranium fission. On December 19, Hahn wrote to Meitner (who was by then living in Sweden) of the barium result, and two days later followed up with a second letter indicating that they were also detecting lanthanum. By chance, Meitner's nephew, physicist Otto Frisch, was then working at Niels Bohr's Institute for Theoretical Physics in Copenhagen. Frisch traveled to Sweden to spend Christmas with his aunt, and they conceived of the fission process around Christmastime, working out an estimate of the energy that could be expected to be released. By this time, Hahn and Strassmann had already submitted their barium paper to the journal *Naturwissenschaften* (Hahn and Strassmann 1939). Otto Hahn was awarded (solely) the 1944 Nobel Prize in Chemistry for the discovery of fission; Meitner and Strassmann did not share in the recognition.

Soon after returning to Copenhagen on New Year's Day 1939, Frisch informed Niels Bohr of the discovery. Bohr was about to depart for a semester at Princeton University, and it is he who carried the word of the discovery to the New World on the same day, January 16, that Meitner and Frisch submitted a paper to *Nature* with their interpretation of the fission process. This was published on February 11 (Meitner and Frisch 1939), by which time the process had been duplicated in a number of laboratories in Europe and America.

Otto Frisch is credited with appropriating the term "fission" from the concept of cell division in biology to describe this newly-discovered phenomenon. He is also credited with being the first person to set up an experiment to deliberately demonstrate it and measure the energy of the fragments, work he did in Copenhagen on Friday, January 13, 1939. After replicating the Hahn & Strassmann uranium results, he also tested thorium. This element proved to act like uranium in that it would fission under bombardment by *fast* neutrons, but at the same time to act *unlike* uranium in that it did not do so at all when bombarded with *slow* neutrons. This asymmetry would catalyze a crucial revelation on the part of Niels Bohr a few weeks later as to which isotope of uranium is responsible for slow-neutron fission. Uranium comprises two isotopes, the "even/even" (in the sense Z/N) one ^{238}U, and the much rarer "even/odd" one ^{235}U, whereas thorium has only one naturally-occurring isotope, ^{232}Th, an "even/even" nuclide. Bohr realized that, as a matter of pure logic, ^{235}U must be responsible for slow-neutron fission, as it is the one "parity" of isotope that thorium does not possess.

The difference in behavior between ^{238}U and ^{235}U under neutron bombardment and how it relates to parity is examined further in Sect. 1.9. In the meantime, we examine in more detail the energetics of the fission process itself.

1.6 Energy Release in Fission

Neutron-induced uranium fission can happen in a multiplicity of ways, with a wide variety of resulting products. Empirically, equal division of the bombarded nucleus is quite unlikely; the most likely mass ratio for the products is about 1.5.

To understand the energy release in fission, consider the splitting of a ^{235}U nucleus into barium and krypton (the Hahn-Strassmann fission-discovery situation), accompanied by the release of three neutrons:

$$\prescript{1}{0}{n} + \prescript{235}{92}{U} \rightarrow \prescript{141}{56}{Ba} + \prescript{92}{36}{Kr} + 3\left(\prescript{1}{0}{n}\right). \tag{1.42}$$

The Δ-values are

$$\begin{cases} \Delta\left(\prescript{1}{0}{n}\right) = 8.071 \\ \Delta\left(\prescript{235}{92}{U}\right) = 40.921 \\ \Delta\left(\prescript{141}{56}{Ba}\right) = -79.726 \\ \Delta\left(\prescript{92}{36}{Kr}\right) = -68.79, \end{cases} \tag{1.43}$$

giving $Q = 173.3$ MeV.

The fission energy latent in a single kilogram of ^{235}U is enormous. With an atomic weight of 235 gr mol^{-1}, 1 kg of ^{235}U comprises about 4.26 mol or 2.56×10^{24} atoms. At 173 MeV/reaction, the potential fission energy amounts to 4.43×10^{32} eV, or 7.1 $\times 10^{13}$ J. Explosion of one *ton* of TNT liberates about 4.2×10^9 J, so the energy released by fission of 1 kg of ^{235}U is equivalent to nearly 17 *kilotons* (kt) of TNT. The explosive yield of the *Little Boy* uranium bomb dropped on Hiroshima has been estimated at about 13 kilotons (Penney et al. 1970), from which we can infer that only some 0.8 kilograms of ^{235}U actually underwent fission. Upon considering that *Little Boy* contained about 53 kg of ^{235}U, we can appreciate that the first fission weapons were rather *inefficient* devices despite their enormous explosive yields. Weapon efficiency is examined in detail in Chap. 2.

In writing the above reaction, it was assumed that three neutrons were released in the process. If one is to have any hope of sustaining a neutron-moderated chain reaction, it is clear that, on average, *at least one* neutron will have to be liberated per fission event. Soon after the discovery of fission, a number of researchers began looking for evidence of these "secondary" neutrons, and proof of their existence was not long in coming. On March 16, 1939, two independent teams at Columbia University submitted letters to *The Physical Review* reporting their discovery: Anderson, Fermi and Hanstein (1939), and Szilard and Zinn (1939). Both groups estimated about two neutrons emitted per each captured. Their papers were published on April 15. In Paris on April 7, Halban et al. (1939) submitted a paper to *Nature* in which they reported 3.5 ± 0.7 neutrons liberated per fission; their paper was published on April 22. The French group, however, made a subtle error in their analysis which resulted in overestimating the number of neutrons per fission; when corrected, their results indicated 2.6 ± 0.6, very much in line with the modern value (Turner 1940; Weart

1979). In an ingenious experiment carried out about the same time and reported in the April 8 edition of *Nature*, Feather (1939) reported that the fission process must take place within a time of no more than about 10^{-13} s. To researchers in the field of nuclear physics, it was apparent in the spring of 1939 that a rapid, extremely energetic uranium-based neutron-initiated-and-maintained chain reaction was at least a theoretical possibility.

1.7   The Bohr-Wheeler Theory of Fission: The Z^2/A Limit Against Spontaneous Fission

Much of the material in this section is adopted from Reed (2003).

Within a few weeks of the discovery of slow-neutron-induced uranium fission, Niels Bohr published a paper in which he argued that of the two then-known isotopes of that element (^{235}U and ^{238}U), it was likely to be nuclei of the lighter, much rarer one that were undergoing fission, whereas nuclei of the heavier isotope would most probably capture any bombarding neutrons and subsequently decay (Bohr 1939). Experimental verification of this prediction came in early 1940 when Alfred Nier separated a small sample of uranium into its constituent isotopes via mass spectroscopy (Nier et al. 1940). In the meantime, Bohr continued with his work on the theory of nuclear fission in collaboration with John Wheeler of Princeton University, efforts which culminated with the publication of a landmark paper in the September 1, 1939, edition of *The Physical Review* (Bohr and Wheeler 1939). In this seminal work, they reported two important discoveries: (i) That there exists a natural limit $Z^2/A \sim 48$ beyond which nuclei are unstable against disintegration by spontaneous fission, and (ii) That in order to induce a nucleus with $Z^2/A < 48$ to fission, it must be supplied with a necessary "activation energy," a quantity also known as the "fission barrier." Uranium isotopes fall into this latter situation.

Before proceeding with any calculations, it is important to point out that the $Z^2/A \sim 48$ limit is not a hard-and-fast one. Uranium is known to fission spontaneously, and has $Z^2/A = (92^2/238) \sim 36$. That this can happen is a consequence of *quantum tunneling*, a wave-mechanical effect discovered independently by George Gamow and the team of Ronald Gurney and Edward Condon in (1928). In this effect, a nucleus can decay by alpha-decay or spontaneous fission even though the process would be energetically forbidden on the basis of classical mechanics. In such decays, the characteristic half-lives can be extremely long; for example, the spontaneous fission half-life for ^{238}U is nearly 10^{16} *years* (see Sect. 4.2). The Bohr and Wheeler theory was purely classical, and sets an upper limit beyond which any nucleus will be expected to essentially instantaneously fission.

Bohr and Wheeler's calculations are extremely challenging, even for advanced physics students. We can, however, get some idea of what they did by invoking some simplifying approximations and by taking some empirical numbers at face value. This section is devoted to an analysis of the issue of the limiting value of Z^2/A against spontaneous fission. Section 1.1.8 examines the energetics of neutrons emitted in the fission process. These two sections set stage for an examination of the fission barrier in Sect. 1.9, which explains why ^{235}U and ^{238}U behave so differently under neutron bombardment. Further analysis of the fission barrier follows in Sect. 1.10, and a numerical simulation of the fission process is examined in Sect. 1.11. A detailed formal treatment of the Bohr and Wheeler analysis is presented in Appendix E.

Bohr and Wheeler modeled nuclei as deformable "liquid drops" whose shapes can be described by a sum of Legendre polynomials configured to conserve volume as they deform. They then considered the total energy of the nucleus to be the sum of two contributions. These are a "surface" energy U_S proportional to the surface area S of the nucleus, and an electrostatic contribution U_C corresponding to its Coulomb self-potential. If a nucleus finds itself deformed from its original spherical shape, U_S will increase due to the consequent increase in surface area, while U_C will decrease as the nuclear charge becomes more spread out. If $(U_S + U_C)_{deformed} < (U_S + U_C)_{original}$, then the nucleus will be unstable against further deformation and potentially eventual fission. The surface-energy term originates in the fact that nucleons near the surface of the nucleus are less strongly bound than those inside, while the Coulomb term arises from mutual repulsion of protons.

The usual textbook approach to establishing the Z^2/A limit is to quote expressions for the surface and Coulomb energies of nuclei modeled as ellipsoids, and then compute the difference in energy between a spherical nucleus and an ellipsoid of the same volume (Fermi 1950). We can obtain an approximate treatment, however, by modeling nuclei as spheres.

Begin with a spherical parent nucleus of radius R_O as shown in Fig. 1.4a. Imagine

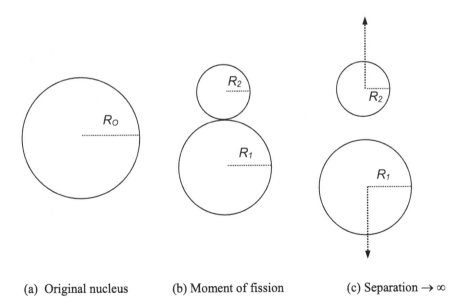

(a) Original nucleus   (b) Moment of fission   (c) Separation → ∞

Fig. 1.4 Schematic illustration of the fission process

that this nucleus splits into two spherical product nuclei of radii R_1 and R_2 as shown in part (b) of the figure; after this, they will repel each other due to the Coulomb force and fly away at high speeds as shown in part (c) of the figure.

Presuming that the density of nuclear matter is constant, conservation of nucleon number demands that volume be conserved:

$$R_O^3 = R_1^3 + R_2^3. \tag{1.44}$$

Define the mass ratio of the fission products as

$$f = \frac{R_1^3}{R_2^3}. \tag{1.45}$$

This ratio could be defined as the inverse of that adopted here, a point to which we shall return below. In terms of this ratio, the radii of the product nuclei are

$$R_1 = R_O \left(\frac{f}{1+f} \right)^{1/3} \tag{1.46}$$

and

$$R_2 = R_O \left(\frac{1}{1+f} \right)^{1/3}. \tag{1.47}$$

Following Bohr and Wheeler, we take the energy of the system at any moment to comprise two contributions: (i) A surface energy proportional to the surface area of the system, and (ii) The Coulombic self-energy of the system. The surface area of the original nucleus is proportional to R_O^2, so its surface energy can be written as

$$U_S^{orig} = a_S R_O^2, \tag{1.48}$$

where the "surface energy coefficient" a_S is to be determined. The surface energy for the fissioned nucleus will be

$$U_S^{fiss} = a_S \left(R_1^2 + R_2^2 \right) = a_S R_O^2 \alpha, \tag{1.49}$$

where

$$\alpha = \frac{f^{2/3} + 1}{(1+f)^{2/3}}. \tag{1.50}$$

For the Coulomb self-energy, begin with the result that the electrostatic self-energy of a charged sphere of radius r is given by

$$U_C^{self} = \left(\frac{4\pi\rho^2}{15\,\varepsilon_o}\right) r^5, \tag{1.51}$$

where ρ is the charge density. In the present case, $\rho = 3\,Z\,e/4\pi\,R_O^3$, where Z is the atomic number of the *parent* nucleus. This gives $4\pi\rho^2/15\,\varepsilon_o = 3\,Z^2 e^2/20\pi\,\varepsilon_o R_O^6$, and hence

$$U_C^{orig} = \left(\frac{3\,e^2\,Z^2}{20\,\pi\,\varepsilon_o\,R_O}\right). \tag{1.52}$$

The electrostatic self-energy of the system at the moment of fission [part (b) Fig. 1.4] is the sum of the self-energies of each of the product nuclei plus the potential energy of the point-charge repulsion between them:

$$U_C^{fiss} = \left(\frac{3\,e^2 Z^2}{20\,\pi\,\varepsilon_o R_O^6}\right)\left(R_1^5 + R_2^5\right) + \frac{q_1\,q_2}{4\,\pi\,\varepsilon_o(R_1 + R_2)}, \tag{1.53}$$

where q_1 and q_2 are the charges of the *product* nuclei. The q's can be expressed in terms of Z and R_O from the presumed-uniform charge density; in terms of this and the mass ratio f, (1.53) reduces to

$$U_C^{fiss} = \left(\frac{3\,e^2 Z^2}{20\,\pi\,\varepsilon_o R_O}\right)(\beta + \gamma), \tag{1.54}$$

where

$$\beta = \frac{f^{5/3} + 1}{(1 + f)^{5/3}} \tag{1.55}$$

and

$$\gamma = \frac{(5/3)\,f}{(1 + f)^{5/3}\left(f^{1/3} + 1\right)}. \tag{1.56}$$

The common factor appearing in (1.52) and (1.54) can be simplified. Empirically, nuclear radii behave as $R \sim a_o A^{1/3}$, where $a_o \sim 1.2$ fm. Incorporating this approximation and substituting values for the constants gives

$$\frac{3\,e^2\,Z^2}{20\,\pi\,\varepsilon_o\,R_O} \sim 0.72 \left(\frac{Z^2}{A^{1/3}}\right)\text{MeV}. \tag{1.57}$$

If the same empirical radius expression is incorporated into the surface-energy expressions, we can absorb the factor of a_o into the definition of a_S and write (1.48) and (1.49) as $U_S^{orig} = a_S A^{2/3}$ and $U_S^{fiss} = a_S A^{2/3}\alpha$; the units of a_S will emerge as MeV.

We can now begin to consider the question of the limiting value of Z^2/A. If the total energy of the two nuclei in the fission circumstance shown in Fig. 1.4b is *less* than that for the original nucleus of Fig. 1.4a, then the system will proceed to fission. That is, spontaneous fission will occur if

$$U_S^{orig} + U_C^{orig} > U_S^{fiss} + U_C^{fiss}. \tag{1.58}$$

Substituting (1.48)–(1.50), (1.52), and (1.54)–(1.57) into (1.58) shows that spontaneous fission will occur for

$$\frac{Z^2}{A} > \frac{a_S (\alpha - 1)}{0.72 (1 - \beta - \gamma)}. \tag{1.59}$$

Estimating the Z^2/A stability limit apparently demands selecting an appropriate mass ratio and knowing the value of a_S. For the latter, we could adopt a value from the semi-empirical mass formula, but it is more satisfying to derive one based on some direct physical grounds. We take up this issue now; the question of an appropriate mass ratio will be addressed shortly.

To calibrate the value of a_S, we appeal to the fact that fission can be induced by slow neutrons with $Q \sim 170$ MeV of energy being liberated. In the present notation this appears as

$$\left(U_S^{orig} + U_C^{orig}\right) - \left(U_S^{\infty} + U_C^{\infty}\right) = Q, \tag{1.60}$$

where U_S^{∞} and U_C^{∞} respectively designate the areal and Coulombic energies of the system when the product nuclei are infinitely far apart. Since the areas of the product nuclei do not change following fission, $U_S^{\infty} = U_S^{fiss}$; see (1.49) and (1.50). U_C^{∞} is given by (1.53) without the point-charge interaction term, that is, (1.54) without the γ term. From these we find

$$a_S = \frac{\left(Q/A^{2/3}\right) - 0.72 \left(Z^2/A\right) (1 - \beta)}{(1 - \alpha)}, \tag{1.61}$$

where A and Z refer to the parent nucleus in the fission reaction, *not* the general Z^2/A spontaneous-fission limit we seek.

Values of a_S derived in this way from a number of fission reactions involving ^{235}U are shown in Table 1.2. The first reaction is representative of the Hahn and Strassmann fission-discovery reaction. The second one is concocted to have the masses of the fission products as 139 and 95, values claimed by Weinberg and Wigner (1958, p. 30) to be the most probable mass yields in slow-neutron fission of ^{235}U. The last two reactions are less probable ones chosen to give a sense of how sensitive a_S is to the choice of calibrating reaction. The mass ratios are those of the fission products, neglecting any neutrons emitted. As one might hope if a_S reflects some fundamental underlying physics, its value is fairly *insensitive* to the choice of calibrating reaction.

Table 1.2 Fission reactions, derived surface energy parameter, and derived spontaneous fission limit

Fission products of $^1n + {}^{235}U$	Q (MeV)	f	a_S (MeV)	$(Z^2/A)_{lim}$
$^{141}_{56}Ba + {}^{92}_{36}Kr + 3\left({}^1_0n\right)$	173.2	1.53	18.3	61.4
$^{139}_{54}Xe + {}^{95}_{38}Sr + 2\left({}^1_0n\right)$	183.6	1.46	17.5	58.6
$^{116}_{46}Pd + {}^{116}_{46}Pd + 4\left({}^1_0n\right)$	177.0	1.00	19.0	63.7
$^{208}_{82}Pb + {}^{26}_{10}Ne + 2\left({}^1_0n\right)$	54.2	8.00	16.3	54.7

The values of a_S derived here are consistent with those quoted in numerical fits of the semi-empirical mass formula, ~ 18 MeV.

With a_S in hand, we face the question of what value of f to use in (1.59) to establish an estimate of the stability limit for Z^2/A against spontaneous fission.

If the limiting value of Z^2/A is to be a matter of fundamental physics, it should in principle be independent of *any* choice for f, although evaluating (1.59) does depend on the choice; in particular, it would make no sense to use the mass ratio for an *induced* reaction used to calibrate a_S to determine a limit against *spontaneous* fission! To resolve this dilemma, recall that f could also have been defined as the inverse of what was adopted in (1.45). The only value of f that is in any sense "unique" is therefore $f = 1$. Indeed, plots of the right side of (1.59) versus f for fixed values of a_S reveals that a minimum always occurs at $f = 1$, symmetric in the sense of $f \to 1/f$ about $f = 1$. To establish a lower limit to the spontaneous-fission condition, let us consequently take (1.59) evaluated at $f = 1$:

$$\left(Z^2/A\right)_{lim} \sim 3.356 a_S. \qquad (1.62)$$

Limiting values of Z^2/A so calculated are given in the last column of Table 1.2; in each case these are based on the a_S values in the preceding column of the Table. While these are somewhat high compared to Bohr and Wheeler value of 48, the agreement is respectable given the simplicity of the model.

Spreadsheet **TwoSphereFission.xls** allows a user to enter mass numbers and Δ-values for reactions like those in Table 1.2; the spreadsheet calculates values for Q, f, α, β, γ, a_S, and the limiting value of Z^2/A.

To close this section, we use this analysis to estimate the value of Z beyond which nuclei will be unstable against spontaneous fission. From data given in the online version of the *Nuclear Wallet Cards*, one finds that there are 352 isotopes that are either permanently stable or have half-lives > 100 years. A plot of A vs. Z for these isotopes can be approximately fit by a power law, as shown in Fig. 1.5; the result is

$$A \sim 1.6864 Z^{1.0870} (r^2 = 0.9965). \qquad (1.63)$$

This fit slightly underestimates $A(Z)$ for heavy nuclei, giving $A \sim 230$ for $Z = 92$, but is sufficiently accurate for our purposes. For a limiting Z^2/A of 60, (1.63) predicts a maximum stable Z of about 157; the Bohr and Wheeler value of 48 gives a maximum Z of about 123.

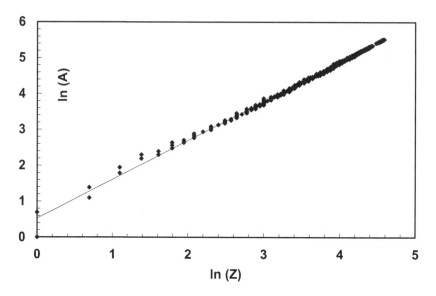

Fig. 1.5 ln(A) versus ln(Z) for 352 nuclides with half-lives > 100 years

1.8   Energy Spectrum of Fission Neutrons

When nuclei fission, they typically emit two or three neutrons. These secondary neutrons are not all of the same energy, however; they exhibit a spectrum of kinetic energies. Knowing the average energy of these "secondary" neutrons is important in understanding the differing fissilities of ^{235}U and ^{238}U, which is analyzed in Sect. 1.9.

According to Hyde (1964), the probability of a neutron being emitted with energy between E and $E+dE$ can be expressed as

$$P(E)\,dE = K\sqrt{E}\,e^{-E/\alpha}dE, \tag{1.64}$$

where K is a normalization constant and α is a fitting parameter; do not confuse this α with that used in the preceding section. For energies measured in MeV, $\alpha \sim 1.29$ MeV in the case of ^{235}U. This distribution is shown in Fig. 1.6.

Formally, (1.64) is mathematically identical to the Maxwell distribution of molecular speeds in statistical mechanics. In this comparison, α would play the role of k_BT, where k_B is Boltzmann's constant and T the absolute temperature. However, there is no underlying theoretical rational for this form in the case of the distribution of neutron energies; it just happens to provide a good empirical fit to measured data.

To determine the normalization factor K, we insist that the sum of the probabilities over all possible energies be unity:

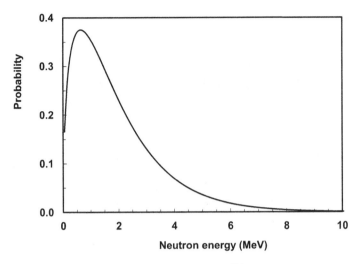

Fig. 1.6 Energy spectrum of neutrons released in fission of ^{235}U

$$\int_0^\infty P(E)\,dE = 1 \Rightarrow K \int_0^\infty \sqrt{E}\,e^{-E/\alpha}\,dE = 1. \tag{1.65}$$

Setting $x = \sqrt{E/\alpha}$ renders this as

$$2K\alpha^{3/2} \int_0^\infty x^2\,e^{-x^2}\,dx = 1. \tag{1.66}$$

This integral evaluates to $\sqrt{\pi}/4$, hence

$$K = \frac{2}{\sqrt{\pi}\,\alpha^{3/2}}. \tag{1.67}$$

To determine the mean neutron energy $\langle E \rangle$, we take a probability-weighted average of E over all possible energies:

$$\langle E \rangle = \int_0^\infty E\,P(E)\,dE = K \int_0^\infty E^{3/2}e^{-E/\alpha}\,dE. \tag{1.68}$$

Again setting $x = \sqrt{E/\alpha}$ renders this in dimensionless form as

$$\langle E \rangle = 2K\alpha^{5/2} \int_{0}^{\infty} x^4 e^{-x^2}\, dx. \tag{1.69}$$

This integral evaluates to $3\sqrt{\pi}/8$; invoking the normalization of (1.67) then gives

$$\langle E \rangle = \frac{3}{2}\alpha. \tag{1.70}$$

Continuing the Maxwell-distribution analogy, recall that the average kinetic energy of particles in a gas at absolute temperature T is $3k_BT/2$. For $\alpha \sim 1.29$ MeV, $\langle E \rangle \sim 1.93$ MeV, or, say, about 2 MeV. From kinetic theory, this is equivalent to a $3k_BT/2$ temperature of $\sim 1.5 \times 10^{10}$ K (see Exercise 1.8 in Appendix H).

In considering the question of why ^{238}U does not make an appropriate material for a weapon (Sect. 1.9), it proves helpful to know what fraction of the secondary neutrons are of energies greater than about 1.4 MeV. For the moment, suffice it to say that the reason for this is that the probability of fissioning ^{238}U nuclei by neutron bombardment is essentially zero for neutrons less energetic than this.

The fraction of neutrons with kinetic energy E greater than some value ε is given by

$$f(E \geq \varepsilon) = \int_{\varepsilon}^{\infty} p(E)dE = K \int_{\varepsilon}^{\infty} \sqrt{E}e^{-E/\alpha}dE. \tag{1.71}$$

This integral can be done by parts (set $x = E/\alpha$), and reduces to

$$f(E \geq \varepsilon) = 1 + \frac{2}{\sqrt{\pi}}ze^{-z^2} + erf(z), \tag{1.72}$$

where $z = \sqrt{\varepsilon/\alpha}$ and where erf is the error function of statistics, which is built into most spreadsheet programs. The behavior of $f(z)$ is shown in Fig. 1.7. For $\varepsilon = 1.4$ MeV and $\alpha = 1.29$ MeV, $z \sim 1.042$, and $f \sim 0.538$. This means that about one-half of the neutrons emitted in the fission of a ^{235}U nucleus would be energetic enough to fission a ^{238}U nucleus. This number will be used not only in the next section but also in Sect. 4.6 in an analysis of the fraction of the yield of the *Trinity* bomb which arose from its ^{238}U tamper. For the present, our interest in this factor of one-half concerns what isotopes can potentially sustain a chain reaction. This is the topic of the next section.

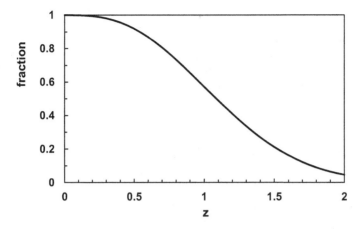

Fig. 1.7 Fraction of fission-liberated neutrons exceeding energy ε MeV , where $z = \sqrt{\varepsilon/\alpha}$, with $\alpha = 1.29$ MeV

1.9 Leaping the Fission Barrier

The isotopes ^{235}U and ^{238}U differ not at all in their chemical properties, but react very differently under neutron bombardment. How can this be? While a detailed treatment of fission is very complex and lies beyond the scope of the present text, we can get some idea of why these isotopes behave so differently by appealing to some energy arguments. The arguments developed in this section are dense; you will need to read them very carefully.

Theory indicates that *any* otherwise stable nucleus can be induced to fission under neutron bombardment. However, any specific isotope possesses a characteristic *fission barrier*. This means that a certain minimum amount of energy has to be supplied to deform the nucleus sufficiently to induce the fission process to proceed. This concept is analogous to the *activation energy* of a chemical reaction; the two terms are used synonymously.

The activation energy can be supplied in two ways: (i) In the form of kinetic energy carried in by the bombarding neutron that initiates the fission, and/or (ii) From "binding" energy liberated when the target nucleus captures the bombarding particle and becomes a different nuclide with its own characteristic mass. Both factors play roles in uranium fission.

The smooth curve in Fig. 1.8 shows theoretically-computed fission barriers in MeV as a function of mass number A; the irregular curve incorporates more sophisticated calculations. Barrier energies vary from a maximum of about 55 MeV for isotopes with $A \sim 90$ down to a few MeV for the heaviest elements such as uranium and plutonium. Half-lives for various modes of decay for elements heavier than Pu tend to be so short as to make them impractical candidates for weapons materials despite their low fission barriers.

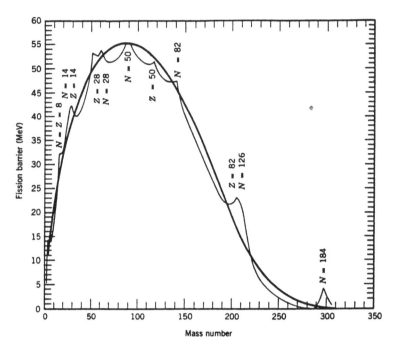

Fig. 1.8 Fission barrier vs. mass number; from Myers and Swiatecki (1966). Reproduced with permission of Elsevier Science

Fission is believed to proceed via formation of an "intermediate" or "compound" nucleus created when the target nucleus captures the incoming neutron. Two cases are relevant for uranium:

$$\mathrm{{}_0^1 n} + \mathrm{{}_{92}^{235}U} \rightarrow \mathrm{{}_{92}^{236}U} \tag{1.73}$$

and

$$\mathrm{{}_0^1 n} + \mathrm{{}_{92}^{238}U} \rightarrow \mathrm{{}_{92}^{239}U}. \tag{1.74}$$

For reaction (1.73), the Δ-values are 8.071, 40.921, and 42.447; the Q-value of this reaction is then 6.545 MeV. For reaction (1.74), the Δ-values are 8.071, 47.310, and 50.575, leading to a Q-value of 4.806 MeV. Now imagine that the bombarding neutrons are "slow," that is, that they bring essentially no kinetic energy into the reactions. ("Fast" and "slow" neutron energies are explored in more detail in Sect. 3.2.) The nucleus of ^{236}U formed in reaction (1.73) will find itself in an *excited state* with an internal energy of about 6.5 MeV, while the ^{239}U nucleus formed in reaction (1.74) will have a like energy of about 4.8 MeV.

Fission barriers for various nuclides are tabulated in Appendix A. For the compound nuclei considered here, ^{236}U and ^{239}U, these are respectively 5.03 and

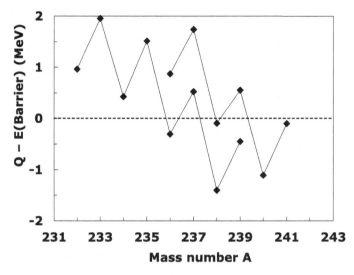

Fig. 1.9 Energy release minus fission barrier for isotopes of uranium (lower line) and plutonium (upper line). If $Q-E_{Barrier} > 0$, an isotope is said to be fissile

6.21 MeV. In the case of ^{236}U, the Q-value exceeds the fission barrier by about 1.5 MeV. Consequently, *any bombarding neutron, no matter how little kinetic energy it has, can induce fission in* ^{235}U. On the other hand, the Q-value of reaction (1.74) falls some 1.4 MeV short of the fission barrier; this is why this particular energy value was investigated in the previous section. To fission ^{238}U by neutron bombardment thus requires input neutrons of at least this amount of energy. ^{235}U is known as a "fissile" nuclide, while ^{238}U is termed "fissionable."

Figure 1.9 shows the situation for various U and Pu isotopes; $Q-E_{Barrier}$ is plotted as a function of target mass number A. The upper line is for Pu isotopes while the lower one is for U isotopes.

From Fig. 1.9, it appears that both ^{232}U and ^{233}U would make good candidates for weapons materials. ^{232}U is untenable, however, as it has a 70-year alpha-decay half-life. For practical purposes, ^{233}U is not convenient as it does not occur naturally and has to be created via neutron capture by thorium in a reactor that is already producing plutonium (Kazimi 2003). ^{234}U has such a low natural abundance as to be of negligible consequence (~ 0.006%), and ^{236}U does not occur naturally at all. ^{237}U has $Q-E_{Barrier} > 0$, but has only a 6.75-day half-life against beta-decay. ^{239}U, the parent of ^{239}Np and ultimately ^{239}Pu, has only a 24 min half-life against beta-decay; this is discussed further below. As physicist David Hafemeister has remarked, nuclear weapons are a fluke of nature: their driving force, ^{235}U, a rare isotope of a relatively rare element, is essentially the only economical path to producing nuclear reactors and weapons (Hafemeister 2014). Pearson (2019) has characterized the possibility of chain reactions as "fortuitous".

As described toward the end of this section, plutonium is "bred" from uranium via neutron capture in a reactor, and one plutonium isotope in particular, ^{239}Pu, makes an excellent fuel for nuclear weapons. Other isotopes of this element are also created in reactors, but those of masses 236, 237, 238 and 241 have such short half-lives against various decay processes as to render them too unstable for use in a weapon (2.87-year alpha-decay, 45-day electron capture, 88-year alpha-decay and 14-year beta-decay, respectively). ^{240}Pu turns out to have such a high spontaneous fission rate that its very presence in a bomb presents a serious danger of causing an uncontrollable premature detonation; this issue is analyzed in Sects. 4.2 and 4.3. ^{239}Pu is the only isotope of that element suitable as a weapons material.

A pattern of alternating high-and-low $Q - E_{Barrier}$ values is evident for both elements in Fig. 1.9. All stable nuclei have lower masses than one would predict on the naïve basis of adding up the masses of their Z protons and A-Z neutrons; the difference goes into binding energy. Nuclear physicists have known for many decades that in this mass-energy sense, so-called even/odd nuclei such as ^{235}U or ^{239}Pu are inherently less stable than even/even nuclei such as ^{238}U; the underlying cause has to do with the way in which nuclear forces act between pairs of nucleons. Expressed qualitatively, even/even nuclei are of even lower mass than the naive mass-addition argument would suggest when compared to what happens with even/odd nuclei. Hence, when an even/odd nucleus such as ^{235}U captures a neutron, it becomes an even/even nucleus of "very" low mass; the mass difference appears as excitation energy via $E = mc^2$. When an even/even nucleus takes in a neutron it liberates mass-energy as well, but not as much as in the even/odd case (we might call the result a "relatively" low mass in comparison); the different $Q - E_{Barrier}$ values are reflected in the jagged lines in Fig. 1.9.

The issue of the unsuitability of ^{238}U as a *weapons* material is, however, more subtle than the above argument lets on. We saw in Sect. 1.8 that the average energy of secondary neutrons liberated in fission of uranium nuclei is about 2 MeV, and that about half of these neutrons have energies greater than the ~1.4 MeV activation energy of the n + ^{238}U $\rightarrow$ ^{239}U reaction. In view of this, it would appear that ^{238}U might make a viable weapons material. Why does it not? The problem turns out to depend on what happens when fast neutrons such as those liberated in fissions encounter ^{238}U nuclei.

The inelastic-scattering cross-section for fission-liberated neutrons against ^{238}U nuclei is about 2.6 barns. In *in*elastic scattering, the kinetic energy of the system is not conserved, and the incoming particle typically suffers a serious reduction in kinetic energy; this is quantified below. In contrast, in *elastic* scattering the kinetic energy of the system is conserved, and in the case of a low-mass particle such as a neutron striking a much higher mass target nucleus, the neutron loses very little kinetic energy. The 2.6-bn figure for inelastic scattering by U-238 is derived from averaging the cross-section over the energy spectrum of the neutrons, a so-called "fission-spectrum average," a concept we will invoke on a number of occasions. On the other hand, the spectrum-averaged fission cross-section for neutrons on ^{238}U is about 0.31 bn. Thus, a fast neutron striking a ^{238}U nucleus is about eight times as likely to be inelastically scattered as it is to induce a fission. Experimentally, neutrons of energy 2.5 MeV inelastically scattering from ^{238}U have their energy reduced to a most probable value of about 0.275 MeV as a result of a *single* scattering; see

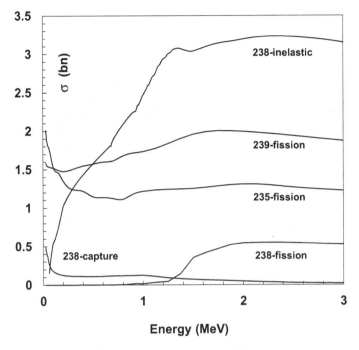

Fig. 1.10 ^{239}Pu, ^{235}U, and ^{238}U fission cross-sections and ^{238}U capture and inelastic-scattering cross-sections as functions of bombarding neutron energy. Adopted from Reed (2008). See also Fig. 1.11

Cranberg and Levin (1956), Fetisov (1957). The vast majority of neutrons striking ^{238}U nuclei will thus promptly be slowed to energies below the fission threshold. The catch is that below about 1 MeV, ^{238}U begins to have a significant non-fission neutron *capture* cross-section, as illustrated in Figs. 1.10 and 1.11.

In short, the non-utility of ^{238}U as a weapons material is due not to a lack of fission cross-section for fast neutrons, but rather to a parasitic combination of inelastic scattering and a fission threshold below which that isotope has an appreciable cross-section for capturing slowed neutrons and removing them from circulation. To aggravate the situation further, the capture cross-section of ^{238}U below about 0.01 MeV is characterized by a dense forest of capture "resonances" with cross-sections of up to thousands of barns, as shown in Fig. 1.11. (The curves in Fig. 1.10 terminate at about 0.03 MeV at the low-energy end. Resonance peaks can be thought of as analogous to how orbital electrons in atoms can be excited to higher energy levels upon capturing photons of just the right energy.) The overall result is a rapid suppression of any putative chain reaction. ^{235}U and ^{239}Pu have cross-sections for inelastic scattering as well, but they differ from ^{238}U in that they have no fission threshold; slowed neutrons will still induce fission in them, and fission will always strongly dominate over capture for them. All of these isotopes also *elastically* scatter neutrons, but this is of no concern here as this process does not significantly degrade the neutrons'

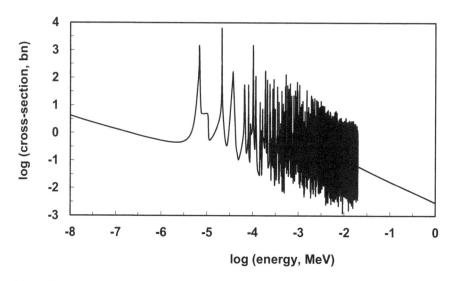

Fig. 1.11 Neutron-capture cross-section for uranium-238. Data from Korean Atomic Energy Research Institute file pendfb7/U238:102. Only about 1% of the available data is plotted here. Many of the resonance capture spikes are so finely spaced that they cannot be resolved

kinetic energies. These cross-section arguments also play a central role in the issue of achieving controlled chain reactions, as is discussed in Sect. 3.1.

To put further understanding to this fast-fission poisoning effect of ^{238}U, consider the following numbers. Suppose (optimistically) that 2-MeV secondary neutrons lose only half their energy due to inelastic scattering. At 1 MeV, the fission cross-section of ^{235}U is about 1.22 bn, while the capture cross-section of ^{238}U is about 0.13 bn. In a sample of natural U, where the ^{238}U:^{235}U abundance ratio is 140:1, capture would consequently dominate fission by a factor of about 15:1 for such neutrons. *The net result is that only ^{235}U can sustain a growing fast-neutron chain reaction, and it is for this reason that this isotope must be isolated from its more populous sister isotope if one aspires to build a uranium fission bomb.* Bomb-grade uranium is defined as 90% pure ^{235}U.

Despite its non-fissility, ^{238}U played a crucial role in the Manhattan Project. The ^{239}U nucleus formed in reaction (1.74) sheds its excess energy in a series of two beta-decays, ultimately giving rise to ^{239}Pu:

$$^{239}_{92}\text{U} \xrightarrow[\text{23.5 min}]{\beta^-} {}^{239}_{93}\text{Np} \xrightarrow[\text{2.36 days}]{\beta^-} {}^{239}_{94}\text{Pu}. \tag{1.75}$$

Like ^{235}U, ^{239}Pu is an even-odd nucleus and was predicted by Bohr and Wheeler to be fissile under slow-neutron bombardment. This is indeed the case. The reaction

$$^{1}_{0}\text{n} + {}^{239}_{94}\text{Pu} \rightarrow {}^{240}_{94}\text{Pu} \tag{1.76}$$

has a Q-value of 6.53 MeV, but the fission barrier of ^{240}Pu is only about 6.0 MeV. It follows that ^{239}Pu will be *fast*-neutron fissile as well, and hence like ^{235}U can serve as the active ingredient in nuclear weapons, although it does prove to have some complications. Plutonium production is explored more fully in Sects. 3.3 and 5.3; the complications are discussed in Sect. 4.2.

To close this section, we make a few follow-up remarks regarding thorium, which was mentioned in the discussion of the discovery of fission in Sect. 1.5. There is only one stable isotope of this element, $^{232}_{90}$Th. This nuclide acts like ^{238}U as far as its inelastic scattering and fissility properties are concerned; the Q-value of the n + ^{232}Th → ^{233}Th reaction is 4.79 MeV, but the fission barrier for ^{233}Th is about 5.5 MeV. Also, the spectrum-averaged fission cross-section of thorium is only about 0.08 barns, so it is only mildly fast-neutron fissile, and is useless as a bomb fuel. It does have value, however, as a potential component of reactor fuel. ^{233}Th decays through protactinium to ^{233}U, which is fissile and so contributes to power production while lessening the amount of plutonium produced—a positive aspect for nuclear non-proliferation efforts.

1.10 A Semi-empirical Look at the Fission Barrier

In this section, we extend the model of the fission process developed in Sect. 1.7 to show how one can "derive" the general trend of fission-barrier energy as a function of mass number shown in Fig. 1.8. This derivation is not rigorous, and will require some interpolation and adoption of a result from Bohr and Wheeler's 1939 paper. Also, as fission barriers are now known to depend in complex ways on nuclear shell effects, pairing corrections, energy levels, and deformation and mass asymmetries, we cannot expect the simple model presented here to capture their detailed behavior. Some elements of this model are adopted from Reed (2020).

We saw in Sect. 1.7 that if a nucleus of mass number A and atomic number Z is modeled as a sphere, its total energy U_E can be expressed as

$$U_E^{orig} = a_S A^{2/3} + a_C \left(\frac{Z^2}{A^{1/3}} \right), \tag{1.77}$$

where a_S and a_C are respectively *surface* and *Coulomb* energy parameters: $a_S \sim$ 18 MeV and $a_C \sim$ 0.72 MeV. Further, if the fissioning nucleus is modeled as two touching spheres of mass ratio f ($f \geq 1$), then the energy of the system at the moment of fission is given by

$$U_E^{fiss} = a_S A^{2/3} \alpha + a_C \left(\frac{Z^2}{A^{1/3}} \right)(\beta + \gamma), \tag{1.78}$$

where

$$\alpha = \frac{f^{2/3} + 1}{(1 + f)^{2/3}}, \tag{1.79}$$

$$\beta = \frac{f^{5/3} + 1}{(1 + f)^{5/3}}, \tag{1.80}$$

and

$$\gamma = \frac{(5/3)\, f}{(1 + f)^{5/3} \left(f^{1/3} + 1\right)}. \tag{1.81}$$

The difference in energy between the fissioned and original configurations is given by

$$\Delta E = U_E^{fiss} - U_E^{orig} = a_S A^{2/3}(\alpha - 1) + a_C \left(\frac{Z^2}{A^{1/3}}\right)(\beta + \gamma - 1). \tag{1.82}$$

Typically, $\Delta E > 0$, that is, there is an energy barrier that inhibits the fission process. The goal here is to look at the behavior of ΔE as a function of the mass number A. For a reason that will become clear in a moment, divide through (1.82) by $a_S A^{2/3}$:

$$\frac{\Delta E}{a_S A^{2/3}} = (\alpha - 1) + \frac{a_C}{a_S}\left(\frac{Z^2}{A}\right)(\beta + \gamma - 1). \tag{1.83}$$

The reason for this manipulation is to set up an expression for ΔE in a form ready to accommodate an important result obtained by Bohr and Wheeler. This is that they were able to prove that the limiting value of Z^2/A against spontaneous fission is given by

$$\left(\frac{Z^2}{A}\right)_{lim} = 2\left(\frac{a_S}{a_C}\right). \tag{1.84}$$

A formal proof of this important result appears in Appendix E. This expression is analogous to (1.62), but is more general as it is entirely independent of the particular shape of the fissioning nucleus. In terms of this limit, we can write (1.83) as

$$\frac{\Delta E}{a_S A^{2/3}} = (\alpha - 1) + 2(\beta + \gamma - 1)\, x, \tag{1.85}$$

where the "fissility parameter" x is defined as

$$x = \left(\frac{Z^2}{A}\right)\Big/ \left(\frac{Z^2}{A}\right)_{lim} \quad (0 < x \le 1). \tag{1.86}$$

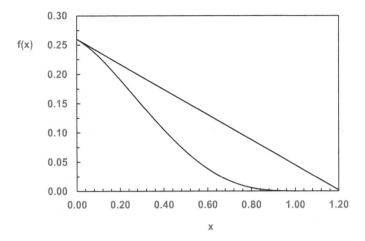

Fig. 1.12 Straight line: fission barrier function $f_{\text{linear}}(x)$ of (1.87). The curved line is the interpolating function of (1.89) and (1.90), configured to give $f_{\text{smooth}}(x) \rightarrow 0$ as $x \rightarrow \sim 1$

Now consider, as did Bohr and Wheeler, fission into equal-mass product nuclei: $f = 1$. In this case we have $\alpha = 1.25992$, $\beta = 0.62996$, $\gamma = 0.26248$, and hence

$$\frac{\Delta E}{a_S A^{2/3}} = f_{\text{linear}}(x) = 0.25992 - 0.21511\, x. \tag{1.87}$$

Equation (1.87) predicts that $\Delta E / a_S A^{2/3}$ will decline linearly with x until it reaches zero at $x = (0.25992/0.21511) = 1.208$; this behavior is shown as the straight line in Fig. 1.12.

That this result predicts a fission barrier of zero for $x > 1$ indicates that a simple "two-sphere" model of fission cannot be an accurate representation of the real shape of a fissioning nucleus; we should have $f(x) \rightarrow 0$ as $x \rightarrow 1$. Presumably, $f(x)$ should have some shape more akin to the curve shown in Fig. 1.12. The precise recipe for the curve shown is elucidated in what follows.

Following Bohr and Wheeler, we develop a plausible interpolating function for $f(x)$. Presuming (as did they) that $f_{\text{linear}}(x)$ accurately models the fission barrier for nuclei with small values of x, we seek an interpolating function that satisfies four criteria:

$$
\begin{aligned}
&\text{(i) } f(x) = (a - 1) \text{ at } x = 0 \\
&\text{(ii) } df/dx = 2(\beta + \gamma - 1) \text{ at } x = 0 \\
&\text{(iii) } f(x) = 0 \text{ at } x = 1 \\
&\text{(iv) } df/dx = 0 \text{ at } x = 1
\end{aligned}
\tag{1.88}
$$

Conditions (i) and (ii) demand that $f(x)$ behave as (1.87) for small values of x. Condition (iii) is the Bohr and Wheeler limiting condition of (1.86), and condition

(iv) ensures that $f(x)$ will approach this limiting condition "smoothly." Apparently, a virtual infinitude of interpolating functions could be conceived. Here we follow Bohr and Wheeler, adopting

$$f_{smooth}(x) = F(1 - x)^3 + B(1 - x)^4, \qquad (1.89)$$

where F and B are constants to be calibrated.

It is not clear from B&W's paper precisely how they were led to this expression. Indeed, in papers published in the spring of 1940, R. D. Present and J. K. Knipp of Purdue University pointed out some algebraic inconsistencies in Bohr and Wheeler's treatment. Purely empirically, however, $f_{smooth}(x)$ has the tremendous advantage that it automatically satisfies conditions (iii) and (iv) above, leaving the coefficients F and B to be determined by conditions (i) and (ii). Thus, *to modify the two-sphere model, I adopt* (1.89), *with the coefficients F and B adjusted so that* $f_{smooth}(x)$ *respects the values of* $f_{linear}(0)$ *and* $(df_{linear}/dx)_0$ *of* (1.87). This gives

$$\left. \begin{array}{l} F = 4(\alpha - 1) \; + \; 2(\beta + \gamma - 1) \\ B = -3\,(\alpha - 1) \; - \; 2(\beta + \gamma - 1) \end{array} \right\}. \qquad (1.90)$$

For $f = 1$, these give $(B, F) = (0.82457, -0.56465)$; this is the smooth curve in Fig. 1.12. With this function that now respects the correct limiting value of Z^2/A, we can write the energy necessary to distort a nucleus to the point of fission as

$$\Delta E = a_S A^{2/3} f_{smooth}(x). \qquad (1.91)$$

To calibrate this model, it is necessary to know the value for ΔE for some nuclide whose x-value is known, in order that we can pin down the limiting value of Z^2/A in (1.86). To do this, I again follow Bohr and Wheeler's lead. They used an estimate of the fission barrier for the compound nucleus ^{239}U (i.e, that formed by neutron capture by ^{238}U) of $\Delta E \sim 6$ MeV. From the table of fission barriers given in Appendix A, the value of this is now estimated to be $\Delta E \sim 6.21$ MeV, so their choice was reasonable. Adopting this value and setting $a_S = 18$ MeV for consistency with Sect. 1.1.7, (1.89)–(1.91) can be solved numerically for x; the result is $x \sim 0.7652$. In (1.86), this gives $(Z^2/A)_{lim} \sim 46.3$, close to Bohr and Wheeler's 1939 value of 47.8. With this, the model is completely calibrated and can be used to predict fission values across the entire range of mass numbers, provided a relationship between Z and A is specified in order that a value of x can be computed for each value of A. For this purpose, if the fit of Eq. (1.63) is reversed, one finds that

$$Z \sim 0.6274\, A^{0.9167}. \qquad (1.92)$$

Figure 1.13 shows the run of ΔE vs. A for $f = 1$ upon assuming (1.92), $a_S = 18$ MeV, and $(Z^2/A)_{lim} = 46.3$ as above.

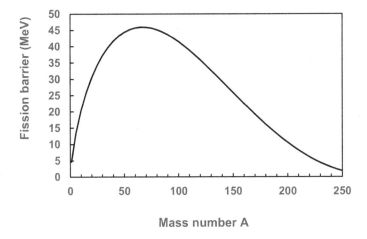

Fig. 1.13 Fission barrier energy of (1.89)–(1.92) for fission-product mass ratio $= 1$. Compare to Fig. 1.8

Figure 1.13 is satisfyingly similar in shape to Fig. 1.8, although with a lower peak value (~ 45 MeV instead of ~ 55). This indicates that (1.89) does not accurately model the run of the fission barrier over all mass numbers, but this is not too surprising as Bohr and Wheeler's purpose was to try to estimate fission barriers for nuclei with x-values similar to that of the ^{239}U compound nucleus in an effort to predict the fissility behavior of heavy nuclei. More detailed fission barrier models can be found in Hyde (1964). For elements lighter than uranium, the fission barrier is too great to be overcome by release of binding energy alone, and it is for this reason that common elements such as aluminum or iron cannot be used to fuel nuclear weapons.

1.11 A Numerical Model of the Fission Process

Material in this section is adopted from Reed (2011).

In Sects. 1.7 and 1.10, the fission process was studied analytically via a simplified "two-sphere" model. While that model does a respectable job of reproducing the overall characteristics of the fission barrier as a function of mass number A (Figs. 1.8 and 1.13), it does have some artificial features and leaves some questions unanswered. Physically, if one imagines a situation midway between panels (a) and (b) of Fig. 1.4, that is, at some point when the nucleus is partially fissioned, there will be a neck joining the product nuclei. Nuclear physicists model surface energies in a manner akin to describing surface tension; a discontinuous curvature would correspond to an infinite force, so the model can be criticized as unrealistic on this aspect alone. More important, perhaps, is the question of the behavior of the energy of the system when the nucleus is only partially fissioned. Does the system energy reach a maximum early on in the separation, or does this occur near the middle or the end of the process? To try to analytically model the intermediate stages of fission is extremely complex: In principle, one would have to examine all possible "deformation trajectories" between an initial spherical configuration and a final two-product configuration, and then isolate the minimum excitation energy in order to determine a minimum barrier energy. It is therefore not surprising that with the rapid development of electronic computers following World War II, numerical simulations came to the fore; models involving up to tenth-order polynomials were published as early as 1947; see Frankel and Metropolis (1947). As a result of this complexity, many texts skip over the details of barrier energetics.

The purpose of this section is to develop a tractable but physically sensible numerical simulation of the fission process. This simulation is based on imagining the nucleus to divide symmetrically into two identical products. A symmetric model does represent some sacrifice of physical reality, but possesses the redeeming feature that the shape of the nucleus can be modeled in such a way that its curvature is never discontinuous.

Suppose that our nucleus begins as a sphere of radius R_O. As a consequence of some disturbance, it begins to distort in a manner that is at all times both axially and equatorially symmetric, as illustrated in Fig. 1.14.

It is assumed that, at any moment, the ends of the distorted nucleus can be modeled as sections of spheres of radius R whose centers are separated by distance $2d$ and which are connected by an equatorial "neck." The outer edge of the neck is taken to be defined by an arc of radius R which is part of a circle that just tangentially touches the spheres which comprise the product nuclei; imagine rotating the figure around

Fig. 1.14 Schematic illustration of a fissioning nucleus

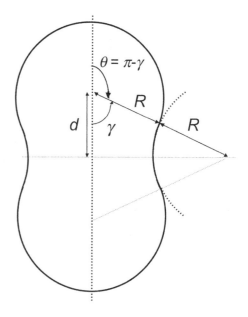

the central vertical axis. Modeling the fission process in this way allows us to avoid
introducing any curvature discontinuities into the shape of the distorted surface.

As the spheres move apart, R must decrease in order to conserve nuclear volume,
as was assumed in Sect. 1.7. The radius of the neck will decrease and eventually reach
zero, at which point the teardrop-shaped products will break contact and fly apart
due to mutual repulsion. A convenient coordinate for parameterizing the process is
the angle γ, which is measured from the polar axis to a line joining the centers of
the upper sphere and the imaginary one that defines the equatorial neck. At the start
of the process, $\gamma = \pi/2$. At any general configuration such as sketched in Fig. 1.14,
the separation of the centers of the emergent spheres is given by

$$2d(\gamma) = 4R(\gamma) \cos \gamma. \tag{1.93}$$

The full height of the equatorial neck at any time is $d(\gamma) = 2R(\gamma) \, cos\gamma$. At
the end of the fission process, the radius of the equatorial neck will have shrunk
to zero, that is, the right-angle triangle given by joining the center of the upper (or
lower) emergent sphere, the center of the nucleus, and the center of the left (or right)
imaginary neck-defining sphere will have a base length R and a hypotenuse of length
$2R$, and hence $\gamma = Sin^{-1}(1/2) = \pi/6$.

Volume and surface areas; volume conservation

In order to formulate an expression for $R(\gamma)$, it is necessary to develop an expression
for the volume of the distorted nucleus and then demand conservation of volume. It
is also necessary to determine the surface area of the nucleus as a function of γ in
order to formulate the surface energy U_S. These issues are taken up first, then the
surface and Coulomb energies are developed.

Figure 1.15 shows a detailed view of the top half of the fissioning nucleus. This
can be imagined to comprise three pieces: a spherical cap where the spherical polar
angle θ (measured toward the equator from the polar axis) runs from zero to $\pi-\gamma$, a
cone of base $R\sin\gamma$ and height $R\cos\gamma$ that nests within the cap, and the equatorial
neck. In what follows, the volumes of these three pieces will be determined separately
and added together to give the overall volume of the nucleus.

Fig. 1.15 Detailed view of
top half of fissioning nucleus

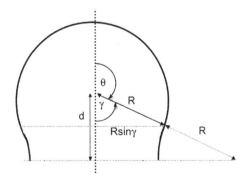

Fig. 1.16 Detailed view of equatorial neck of fissioning nucleus

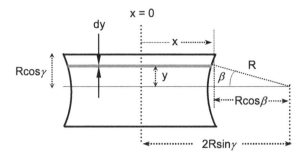

The volumes of the spherical cap and the cone are fairly straightforward. An element of volume in spherical coordinates is given by $dV = r^2 \sin\theta \, dr \, d\theta d\phi$, so the volume of the *two* end caps is

$$V_{caps} = 2 \int_{r=0}^{R} \int_{\theta=0}^{\pi-\gamma} \int_{\phi=0}^{2\pi} r^2 \sin\theta \, dr \, d\theta \, d\phi = \frac{4\pi}{3} R^3 (1 + \cos\gamma). \tag{1.94}$$

The volume of a right circular cone is 1/3 times the area of its base times its height; accounting for both the top and bottom cones gives

$$V_{cones} = \frac{2\pi}{3} R^3 \sin^2\gamma \cos\gamma. \tag{1.95}$$

The volume of the equatorial neck is somewhat more complicated. Figure 1.16 shows the neck in more detail. The angle β lies between the equatorial plane and a radial line from the center of the imaginary neck-defining sphere to a point on the neck.

The shaded strip in the figure represents a disk of thickness dy and radius x that is located at height y above the mid-plane of the nucleus. Since the distance from the center of the nucleus to the center of the (imaginary) sphere that is used to define the neck is $2R\sin\gamma$ (Fig. 1.14), the radius of the disk is

$$x = R(2\sin\gamma - \cos\beta). \tag{1.96}$$

The height of the disk above the equatorial plane is $y = R\sin\beta$, so the thickness of the disk must be

$$dy = R\cos\beta \, d\beta. \tag{1.97}$$

The volume of the disk is $\pi x^2 dy$. The limits on β are zero to $\pi/2 - \gamma$; on taking both halves of the neck into account, its volume is

$$V_{neck} = 2\pi R^3 \int_0^{\pi/2-\gamma} \cos\beta(2\sin\gamma - \cos\beta)^2 d\beta. \tag{1.98}$$

This integral is tedious but straightforward, and gives

$$V_{neck} = \frac{2\pi R^3}{3}\left\{13\sin^2\gamma\cos\gamma + 2\cos\gamma - 3\sin\gamma\left[\pi - 2\gamma + \sin(2\gamma)\right]\right\}. \tag{1.99}$$

On adding (1.94), (1.95), and (1.99), the volume of the partially-fissioned nucleus can be written as

$$V = \frac{4}{3}\pi R^3 f_V(\gamma), \tag{1.100}$$

where

$$f_V(\gamma) = 1 + 7\sin^2\gamma\cos\gamma + 2\cos\gamma - \frac{3}{2}\sin\gamma\left[\pi - 2\gamma + \sin(2\gamma)\right]. \tag{1.101}$$

The original nucleus was of volume $4\pi R_O^3/3$, so demanding conservation of volume during the fission process gives the radius R at any time in terms of R_O and γ as

$$R(\gamma) = f_V^{-1/3}(\gamma)R_O. \tag{1.102}$$

Calculation of the surface area of the deformed nucleus proceeds similarly. The element of surface area in spherical coordinates is $dS = r^2 \sin\theta\, d\theta\, d\phi$. The surface area of the two end caps is

$$S_{caps} = 2R^2 \int_{\theta=0}^{\pi-\gamma} \int_{\phi=0}^{2\pi} \sin\theta\, d\theta\, d\phi = 4\pi R^2(1 + \cos\gamma). \tag{1.103}$$

The cones contribute no surface area as they are embedded within the caps. As for the neck, look again to Fig. 1.16. An element of arc length along the edge of the neck for angle $d\beta$ will be $Rd\beta$. The area of the edge of the narrow disk must then be $2\pi \times R\, d\beta$. Hence, for the entire neck, we have, with (1.96) for x,

$$S_{neck} = 4\pi R^2 \int_0^{\pi/2-\gamma} (2\sin\gamma - \cos\beta)d\beta. \tag{1.104}$$

This evaluates to

$$S_{neck} = 4\pi R^2 \{2 \sin \gamma (\pi/2 - \gamma) - \cos \gamma\}. \qquad (1.105)$$

[Note: In the paper from which this material is adopted, Reed (2011), S_{neck} is given incorrectly.] Gathering (1.103) and (1.105), we can write the surface area of the distorted nucleus as

$$S = 4\pi R^2(\gamma) f_S(\gamma) \qquad (1.106)$$

where

$$f_S(\gamma) = 1 + 2 \sin \gamma (\pi/2 - \gamma). \qquad (1.107)$$

Two sundry results are also listed here. At the moment when the equatorial neck has shrunk to zero radius ($\gamma = \pi/6$), the radius of each spherical end-cap is

$$R_{fiss} = \left(\frac{2}{3\sqrt{3} + 2 - \pi}\right)^{1/3} R_O \sim 0.790 R_O. \qquad (1.108)$$

The separation of the centers at this time will be $2d_{fiss} = 2\sqrt{3} R_{fiss} \sim 2.737 R_O$, and the full height of the distorted nucleus will be $h = 2R_{fiss} + 2d_{fiss}$, which reduces to

$$h = 2(1 + \sqrt{3}) \left(\frac{2}{3\sqrt{3} + 2 - \pi}\right)^{1/3} R_O \sim 4.317 R_O. \qquad (1.109)$$

Surface and Coulomb Energies

As in Sect. 1.7, the surface energy of the deformed nucleus is assumed to be directly proportional to its surface area S at any moment. Again invoke the empirical result that nuclear radii behave as $R \sim a_o A^{1/3}$, where $a_o \sim 1.2$ fm. This result along with (1.102) and (1.106) gives the surface area as

$$S = 4\pi a_o^2 A^{2/3} \Sigma_S, \qquad (1.110)$$

where

$$\Sigma_S = f_S(\gamma) f_V^{-2/3}(\gamma). \qquad (1.111)$$

Σ_S is purely a function of the emergence angle γ. If we introduce Ω as the value of the conversion factor from surface area in square meters to energy in MeV, the product $4\pi a_o^2 \Omega$ is the surface energy parameter $a_S \sim 18$ MeV of Sects. 1.7 and 1.10. We can then write the surface energy as

$$U_S = a_S A^{2/3} \Sigma_S. \text{ (MeV)} \tag{1.112}$$

The Coulomb energy of the deformed nucleus is determined by direct numerical integration. The fissioning nucleus is imagined to be situated within a lattice of volume elements, and the self-energy is computed by adding up the Coulomb energies of all pairs of elements that find themselves within the nucleus. While this procedure is conceptually straightforward, there are a number of practical issues to deal with: Ensuring that the number of volume elements is sufficiently large to obtain reasonable accuracy without rendering the computation prohibitively time-consuming, writing a program to take advantage of the symmetry of the situation to minimize the number of computations, and adopting a convenient system of units.

Imagine surrounding the configuration of Fig. 1.14 with a cylindrical lattice of N cells comprising N_Z vertical layers, with each layer consisting of N_R radial rings and N_ϕ angular wedges: $N = N_Z N_R N_\phi$. The lattice remains of constant volume throughout the integration (as do the individual cells), and is set to have a radius equal to that of the initial undeformed nucleus (R_O) and a height great enough to accommodate the just-fissioned system, $h = \eta R_O$, where η is the factor 4.317... of (1.109). The bottom of the lower sphere is taken at every step in the calculation to be sitting at $(x, y) = (0, 0)$. R_O is used as the unit of distance. In the program set up to carry out this computation, the radial rings are configured to become thinner with increasing radius in order that all cells are of the same volume; the radius of the i'th ring (in units of R_O) is given by $r_i = (i/N_R)^{1/2}$.

The volume of any one of the lattice cells is $\pi \eta R_O^3 / N$. If the nucleus contains Z protons, its charge density will be $\rho = 3eZ / 4\pi R_O^3$, and so the charge contained in a cell that lies within the nucleus will be $Q = 3eZ\eta/4N$. The Coulomb potential between any two "occupied" cells (i, j) separated by distance r_{ij} will be $Q_i Q_j / 4\pi \varepsilon_o r_{ij}$. On again setting $R_O = a_o A^{1/3}$, writing $r_{ij} = R_O d_{ij}$, and summing over all pairs of cells, the total Coulombic potential energy emerges as

$$U_C = \left(\frac{15}{16}\right) a_C \frac{Z^2}{A^{1/3}} \left(\frac{\eta}{N}\right)^2 \Sigma_C, \tag{1.113}$$

where a_C is the Bohr-Wheeler Coulomb energy parameter of Sect. 1.7,

$$a_C = \frac{3e^2}{20\pi \varepsilon_o a_o} \sim 0.72 \text{MeV}, \tag{1.114}$$

and where

$$\Sigma_C = \sum_{i=1}^{N-1} \sum_{j=i+1}^{N} \frac{\delta_i \delta_j}{d_{ij}}. \tag{1.115}$$

In (1.115), $\delta_k = 1$ if lattice cell k lies within the nucleus, and zero if it does not. Like Σ_S, Σ_C is purely a function of the emergence angle γ. Combining (1.111)–(1.115), the total (surface + Coulomb) configuration energy of the nucleus becomes

$$U_{Total} = a_S A^{2/3} \left\{ \Sigma_S + \left(\frac{15}{16} \right) \left(\frac{a_C}{a_S} \right) \left(\frac{Z^2}{A} \right) \left(\frac{\eta}{N} \right)^2 \Sigma_C \right\}. \qquad (1.116)$$

With $a_S = 18$ MeV and $a_o = 1.2$ fm, $15a_C/16a_S \sim 0.0375$. These values are assumed in what follows.

1.12 Results

The publication from which this material is adopted, Reed (2011), describes programs for carrying out the above calculations. These programs need only be run once, after which the run of total energy for any (A, Z) as a function of γ can be obtained by scaling Σ_S and Σ_C according as (1.116). The programs use a lattice comprising two million cells: $(N_Z, N_R, N_\phi) = (200, 100, 100)$; this choice was found to provide both reasonably expedient run times and sensible accuracy. To minimize computation time, the programs assume that the nucleus is axially symmetric, although not equatorially symmetric. The results of the programs are listings of Σ_S and Σ_C as functions of γ; these have been collected into a spreadsheet, **ActivationCurve.xls**, which can be employed to evaluate (1.116) for any choice of (A, Z).

Figure 1.17 shows the deformation energy, that is, the change in total configuration energy in the sense (deformed nucleus minus original nucleus) versus d/R_O for nuclei of uranium $[(A, Z) = (235, 92)]$ and zirconium $[(A, Z) = (90, 40)]$. The value of d/R_O at the moment of fission is 1.369; see the discussion following (1.108) above.

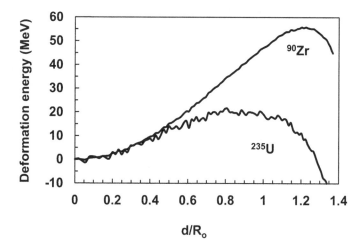

Fig. 1.17 Deformation energy curves for ^{90}Zr (top curve) and ^{235}U (bottom curve)

Zirconium is used as an example here because, for real nuclei, more sophisticated models reveal that fission barriers peak at a value of ~55 MeV for nuclei with A ~90 (Fig. 1.8). The simulation-computed maximum of ~55.9 MeV for zirconium is in very good agreement with this value, although that for uranium, ~22 MeV, is high compared to the true value of ~6 MeV. In the case of uranium, the maximum distortion energy is reached at about the middle of the fission process, whereas for zirconium the maximum occurs at near the end of the process. At the moment of fission, the value of ΔE for uranium has dropped to about -23 MeV.

The numerical precision achieved with this model can be judged by examining how well it reproduces the configuration energy for the initial undeformed nucleus. This can be computed analytically as [see (1.77)]

$$U_{start} = a_S A^{2/3} \left\{ 1 + \left(\frac{a_C}{a_S} \right) \left(\frac{Z^2}{A} \right) \right\}. \tag{1.117}$$

For $(A, Z) = (235, 92)$ and $(a_C, a_S) = (0.72\ \text{MeV}, 18\ \text{MeV})$, this gives $U_{start} = 1673.0$ MeV; the simulation gives 1673.74 MeV, only 0.04% high.

Proceeding to some other nuclides, running $(A, Z) = (208, 82)$ to simulate the most common stable isotope of lead gives a barrier of ~32 MeV, whereas Fig. 1.8 indicates ~25 MeV. Toward the lighter end of the periodic table, choosing $(A, Z) = (50, 22)$, an isotope of titanium, gives a barrier of ~ 53 MeV, in close agreement with Fig. 1.8. For neon with $(A, Z) = (20, 10)$, the simulation gives a barrier of ~34 MeV. This is about twice the value indicated in Fig. 1.8, but barrier energies rise very rapidly at low mass numbers. At the other extreme, the heavy synthetic element darmstadtium with $(A, Z) = (270, 110)$ has a calculated barrier of only about 8 MeV before its ΔE curve drops precipitously to ~ -80 MeV at the moment of fission. Such a nuclide is thus essentially spontaneously fissile, as one might expect for its Z^2/A value of ~45.

Figure 1.18 shows the run of maximum computed ΔE vs. mass number A. This was formed by computing $\Delta E(A)$ for values of $A = 10, 15, 20, \ldots 300$ and searching for the maximum value of ΔE for each A; the curve is interpolated.

The model predicts somewhat high values of ΔE at both low and high values of A and somewhat low ones for intermediate values, but it does successfully reproduce the overall trend of $\Delta E(A)$. In computing these maximum ΔE values, it was assumed that the atomic number Z for a given value of A could be modeled as in (1.92), $Z \sim 0.627 A^{0.917}$ ($1 \leq Z \leq 98$).

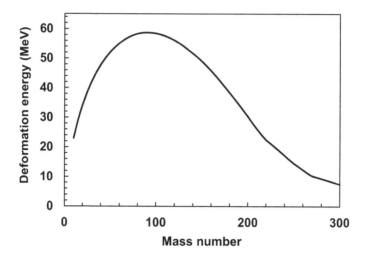

Fig. 1.18 Fission barrier vs. mass number: Results of numerical simulation

References

Amaldi, E.: From the discovery of the neutron to the discovery of nuclear fission. Phys. Rep. **111**(1–4), 1–332 (1984)

Anderson, H.L., Fermi, E., Hanstein, H.B.: Production of neutrons in uranium bombarded by neutrons. Phys. Rev. **55**, 797–798 (1939)

Bohr, N.: Resonance in uranium and thorium disintegrations and the phenomenon of nuclear fission. Phys. Rev. **55**, 418–419 (1939)

Bohr, N., Wheeler, J.A.: The mechanism of nuclear fission. Phys. Rev. **56**, 426–450 (1939)

Brown, A.: The Neutron and the Bomb: A Biography of Sir James Chadwick. Oxford University Press, Oxford (1997)

Chadwick, J.: Possible existence of a neutron. Nature **129**, 312 (1932a)

Chadwick, J.: The existence of a neutron. Proc. R. Soc. London. **A136**, 692–708 (1932b)

Coster-Mullen, J.: Atom Bombs: The Top Secret Inside Story of Little Boy and Fat Man. Coster-Mullen, Waukesha, WI (2016)

Cranberg, L., Levin, J.S.: Neutron scattering at 2.45 Mev by a Time-of-Flight Method. Phys. Rev. **103**(2), 343–352 (1956)

Feather, N.: The time involved in the process of nuclear fission. Nature **143**, 597–598 (1939)

Fermi, E.: Nuclear Physics. University of Chicago Press, Chicago (1950; 1974 paperback reprint), p. 165

Fetisov, N.I.: Spectra of neutrons inelastically scattered on ^{238}U. At. Energ. **3**, 995–998 (1957)

Frankel, S., Metropolis, N.: Calculations in the liquid-drop model of fission. Phys. Rev. **72**, 914–925 (1947)

Gamow, G.: Zur Quantentheorie des Atomkernes (On Quantum Theory of Atomic Nuclei). Z. Angew. Phys. **51**, 204 (1928)

Gurney, R.W., Condon, E.U.: Quantum mechanics and radioactive disintegration. Nature **122**(3073), 439 (1928)

Hafemeister, D.: Physics of Societal Issues, p. 1. Springer International Publishing, Switzerland (2014)

Hahn, O., Strassmann, F.: Über den Nachweis und das Verhalten der bei der Bestrahlung des Urans mittels Neutronen entstehenden Erdalkalimetalle. Naturwissenschaften **27**, 11–15 (1939). An

English translation appears in Graetzer, H. G.: Discovery of Nuclear Fission. Am. J. Phys. **32**, 9–15 (1964). In English, the title is "Concerning the Existence of Alkaline Earth Metals Resulting from Neutron Irradiation of Uranium."

Hyde, E.K.: The Nuclear Properties of the Heavy Elements III. Fission Phenomena. Prentice-Hall, Englewood Cliffs (1964)

Kazimi, M.S.: Thorium fuel for nuclear energy. Am. Sci. **91**, 408–415 (2003)

Meitner, L., Frisch, O.: Disintegration of uranium by neutrons: a new type of nuclear reaction. Nature **143**, 239–240 (1939)

Myers, W., Swiatecki, W.: Nuclear masses and deformations. Nuclear Physics **81**, 1–60 (1966)

Nier, A.O., Booth, E.T., Dunning, J.R., Grosse, A.V.: Nuclear fission of separated uranium isotopes. Phys. Rev. **57**, 546 (1940)

Pearson, J.M.: The fortuitous chain reaction. Am. J. Phys. **87**(4), 264–269 (2019)

Penney, W., Samuels, D.E.J., Scorgie, G.C.: The nuclear explosive yields at Hiroshima and Nagasaki. Phil. Trans. Roy. Soc. London. **A266**(1177), 357–424 (1970)

Present, R.D., Knipp, J.K.: On the dynamics of complex fission. Phys. Rev. **57**, 751 (1940a)

Present, R.D., Knipp, J.K.: On the dynamics of complex fission. Phys. Rev. **57**, 1188–1189 (1940b)

Reed, B.C.: Simple derivation of the Bohr-Wheeler spontaneous fission limit. Am. J. Phys. **71**, 258–260 (2003)

Reed, B.C.: Chadwick and the Discovery of the Neutron. Society of Physics Students Observer, **XXXIX**(1), 1–7 (2007)

Reed, B.C.: A graphical examination of uranium and plutonium fissility. J. Chem. Educ. **85**(3), 446–450 (2008)

Reed, B.C.: A desktop-computer simulation for exploring the fission barrier. Nat. Sci. **3**(4), 323–327 (2011)

Reed, B.C.: The History and Science of the Manhattan Project, 2nd edn. Springer, Berlin (2019)

Reed, B.C.: A pedagogical reconstruction of Bohr and Wheeler's fission-barrier graph. Eur. J. Phys. In press (2020)

Rhodes, R.: The Making of the Atomic Bomb. Simon and Schuster, New York (1986)

Rutherford, E., Soddy, F.: Radioactive change. Phil. Mag. series 6, **v**, 576–591 (1903)

Rutherford, E.: Collision of α particles with light atoms. IV. An Anomalous Effect in Nitrogen. Phil. Mag. series 6, **xxxvii**, 581–587 (1919)

Sime, R.L.: Lise Meitner: A Life in Science. University of California Press, Berkeley (1996)

Szilard, L., Zinn, W.H.: Instantaneous emission of fast neutrons in the interaction of slow neutrons with uranium. Phys. Rev. **55**, 799–800 (1939)

Turner, L.A.: Secondary neutrons from uranium. Phys. Rev. **57**, 334 (1940)

von Halban, H., Joliot, F., Kowarski, L.: Number of neutrons liberated in the nuclear fission of uranium. Nature **143**, 680 (1939)

Weart, S.R.: Scientists in Power. Harvard University Press, Cambridge (1979). See pp. 85–86 and 303–304

Weinberg, A.M., Wigner, E.P.: The Physical Theory of Neutron Chain Reactors. University of Chicago Press, Chicago (1958)

Chapter 2
Critical Mass, Efficiency, and Yield

This chapter forms the heart of this book. Every gram of enriched uranium or synthe-sized plutonium produced in the Manhattan Project was obtained at great cost and with great difficulty, so estimating the amount of fissile material needed to make a workable nuclear weapon—the so-called critical mass—was a crucial issue for the developers of *Little Boy* and *Fat Man*. Equally important was to estimate what efficiency one might expect for a bomb. For various reasons, not all of the mate-rial in a bomb core undergoes fission during a nuclear explosion; if the expected efficiency were to prove so low that one might just as well use a few conventional bombs to achieve the same energy release, there would be no point in taking on the massive engineering challenges involved in making nuclear weapons. In this chapter we investigate these issues.

The concept of critical mass involves two competing effects. As nuclei fission, they emit secondary neutrons. A fundamental empirical law of nuclear physics, derived in Sect. 2.1, shows that while some neutrons will cause other fissions, the remainder will reach the surface of the mass and escape. If, however, more than one neutron is emitted per fission, we can afford to let some escape since only one is required to initiate a subsequent fission. For a small sample of material the escape probability is high; as the size of the sample increases, the escape probability declines and at some point will reach a value such that the number of neutrons that fail to escape will be sufficient to fission every nucleus in the mass—in theory, at least. Thus, there is a minimum size (hence mass) of material for which every nucleus will in principle fission even while some neutrons escape.

The above description of critical mass should be regarded as being only quali-tative. Technically, the important issue is known as *criticality*. Criticality is said to obtain when the number of free neutrons inside a bomb core is increasing with time. A full understanding of criticality demands familiarity with time-dependent diffu-sion theory. Application of diffusion theory to this problem requires understanding a concept known as the *mean free path* (MFP) for neutron travel, so this is also developed in Sect. 2.1. Section 2.2 takes up a time-dependent diffusion theory treat-ment of criticality for the simplest possible configuration: a sphere of fissile material

© The Author(s), under exclusive license to Springer Nature Switzerland AG 2021
B. C. Reed, *The Physics of the Manhattan Project*,
https://doi.org/10.1007/978-3-030-61373-0_2

not surrounded by a *tamper*, that is, a so-called *bare* or *naked* core. A tamper is a heavy metal casing which enhances weapon efficiency in two ways: By reflecting escaped neutrons back into the core and hence giving them fresh chances at causing fissions, and by briefly retarding the violent expansion of the core in order to give the chain reaction more time over which to operate. Providing a tamper can significantly increase the efficiency of a weapon at a very low cost, and tampers were employed in both the *Little Boy* and *Fat Man* bombs of the Manhattan Project. Tamped criticality is taken up in Sect. 2.3. Another modification to bomb design which was contemplated during the Manhattan Project but not utilized is that of constructing a *composite core*, that is, one comprising nested shells of two different fissile materials. The point of this is to optimize bomb production: If, for example, you are producing ^{235}U and ^{239}Pu at the same rate, you can in principle produce bombs more steadily if they are designed with cores having equal amounts of each material, as opposed to waiting until you have enough for one of each type. This description is loose in that the exact mass ratio of the two materials that is necessary to achieve criticality will depend on their individual fissility characteristics, but makes the point that bomb productivity can be improved in this way. Tamped composite cores are analyzed in Sect. 2.4. Sections 2.5 and 2.6 take up the issue of bomb efficiency and yield through analytic approximations and a numerical simulation. Section 2.7 presents an alternate treatment of untamped criticality that has an interesting historical connection. Section 2.8, which is new in this edition of this book, presents several approximate methods for analyzing critical mass and bomb efficiency. The point of these less rigorous analyses is to provide treatments that can be used to make quick estimates or which might be suitable for classroom discussions when time is at a premium. It must be remembered, however, that they are approximate treatments, and that for more precise answers one should return to the fuller diffusion-theory treatments. Finally, Sect. 2.9 presents an approximate treatment of criticality for cylindrical bomb cores.

For readers interested in further sources, an excellent account of the concept of critical mass appears in Logan (1996); see also Bernstein (2002).

2.1 Cross-Sections, Mean Free Path, and the Diffusion Equation

See Fig. 2.1. A thin slab of material of thickness s (ideally, one atomic layer) and cross-sectional area Σ is bombarded by incoming neutrons at a rate R_o neutrons per square meter per second. Such an areal-specified rate is known as the neutron flux.

Let the bulk density of the material be ρ gr cm^{-3}. In nuclear reaction calculations, however, density is usually expressed as a *number density* of nuclei in the material, that is, as the number of nuclei per cubic meter. In terms of ρ, this is given by

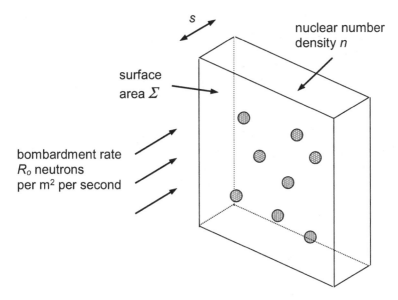

Fig. 2.1 Neutrons penetrating a thin target foil

$$n = 10^6 \left(\frac{\rho N_A}{A} \right), \tag{2.1}$$

where N_A is Avogadro's number and A is the atomic weight of the material in grams per mole; the factor of 10^6 arises from converting cm^3 to m^3.

Assume that each nucleus presents a total reaction cross-section of σ square meters to the incoming neutrons. Cross-sections are usually measured in barns (bn, or just b), where 1 bn $= 10^{-28}$ m^2, a value characteristic of the physical sizes of nuclei. The first question we address is: "How many reactions will occur per second as a consequence of the bombardment rate R_o?" The volume of the slab is Σs, hence the number of nuclei contained in it will be Σsn. If each nucleus presents an effective cross-sectional area σ to the incoming neutrons, then the total area presented by all nuclei would be $\Sigma sn\sigma$. The *fraction* of the surface area of the slab that is available for reactions to occur is then $(\Sigma sn\sigma/\Sigma) = sn\sigma$. The rate of reactions R (reactions per second) can then sensibly be assumed to be the rate at which incoming particles bombard the surface area of the slab, times the fraction of the surface area available for reactions:

$$\left(\begin{array}{c} reactions\ per \\ second \end{array} \right) = \left(\begin{array}{c} incident\ neutron \\ flux\ per\ second \end{array} \right) \left(\begin{array}{c} fraction\ of\ surface\ area \\ occupied\ by\ cross - section \end{array} \right),$$

or

$$R = (R_o \Sigma)(s\,n\,\sigma). \tag{2.2}$$

The concept of a reaction rate will be used extensively in Sects. 3.3 and 4.5 for estimating the rates of production of plutonium and polonium within a reactor. For the present purpose, our interest is in the probability that neutron will pass through the layer *without* striking a nucleus. To approach this, it is easier to begin with the probability that a neutron will in fact precipitate a reaction:

$$P_{react} = \frac{\left(\begin{array}{c} reactions \\ per\ second \end{array}\right)}{\left(\begin{array}{c} incident\ neutron\ flux \\ per\ second \end{array}\right)} = \frac{(R_o\Sigma)(s\,n\,\sigma)}{(R_o\Sigma)} = s\,n\,\sigma. \qquad (2.3)$$

The probability that a neutron will pass through the slab to escape out the back side will then be:

$$P_{escape} = 1 - P_{react} = 1 - s\,n\,\sigma. \qquad (2.4)$$

Now consider a block of material of macroscopic thickness x. As shown in Fig. 2.2, we can imagine this comprising a large number of thin slabs each of thickness s placed back-to-back.

The number of slabs is x/s. If N_o neutrons are incident on the left side of the block, the number that would survive to emerge from the first thin slab would be N_oP, where P is the escape probability in (2.4). These neutrons are then incident on the second slab, and the number that would emerge unscathed from that passage would be $(N_oP)P = N_oP^2$. These neutrons would then strike the third slab, and so on. The number that survive passage through the entire block to escape from the right side would be $N_oP^{x/s}$, or

$$N_{esc} = N_o(1 - s\,n\,\sigma)^{x/s}. \qquad (2.5)$$

Define $z = -sn\sigma$. The number of neutrons that escape can then be written as

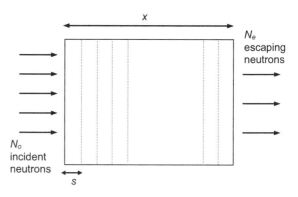

Fig. 2.2 Neutrons penetrating a thick target

$$N_{esc} = N_o(1 + z)^{-\sigma n x/z} = N_o\big[(1 + z)^{1/z}\big]^{-\sigma n x}. \qquad (2.6)$$

Now, ideally, s is very small, which means that $z \to 0$. The definition of the base of the natural logarithms, e, is $e = \lim_{z \to 0}(1 + z)^{1/z}$, so we have

$$N_{esc} = N_o e^{-\sigma n x},$$

or

$$P_{\substack{direct \\ escape}} = \frac{N_{esc}}{N_o} = e^{-\sigma n x}. \qquad (2.7)$$

Equation (2.7) is the fundamental neutron escape probability law. In words, it says that the probability that a bombarding neutron will pass through a slab of material of thickness x depends exponentially on the product of x, the number density of nuclei in the slab, and the reaction cross-section of the nuclei to incoming neutrons. If $\sigma = 0$, all of the incident particles will pass through unscathed. If $(\sigma n x) \to \infty$, *none* of the incident particles will make it through.

In practice, (2.7) is used to experimentally establish values for cross-sections by bombarding a slab of material with a known number of incident particles and then seeing how many emerge from the other side; think of (2.7) as effectively *defining* σ. Due to quantum-mechanical effects, the cross-section is *not* the geometric area of a nucleus.

The total cross section in mind here can be broken down into a sum of cross-sections for individual processes such as fission, elastic scattering, inelastic scattering, non-fission capture, etc.:

$$\sigma_{total} = \sigma_{fission} + \sigma_{\substack{elastic \\ scatter}} + \sigma_{\substack{inelastic \\ scatter}} + \sigma_{capture} + \cdots \qquad (2.8)$$

In practice, cross-sections can depend very sensitively on the energy of the incoming neutrons, as was seen in Figs. 1.10 and 1.11. Figure 2.3 shows another example, the variation of the fission cross-section for ^{235}U under neutron bombardment for neutrons in the energy range 1–10 eV; see also Fig. 3.1, which shows the fission cross-section for ^{235}U across many orders of magnitude of bombarding-neutron energy. These energy-dependences play a crucial role in the difference between how nuclear reactors and nuclear weapons function.

A very important result that derives from this escape-probability law is an expression for the *average* distance that an incident neutron will penetrate into the slab before being involved in a reaction. Look at Fig. 2.4, where we now have a slab of thickness L and where x is a coordinate for any position within the slab. Imagine also a small slice of thickness dx whose front edge is located at position x.

From (2.7), the probability that a neutron will penetrate through the entire slab to emerge from the face at $x = L$ is $P_{emerge} = e^{-\sigma n L}$. This means that the probability that a neutron will be involved in a reaction and *not* travel through to the face at

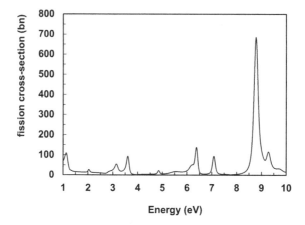

Fig. 2.3 Cross-section for the ^{235}U(n, f) reaction over the energy range 1-10 eV. At 0.01 eV (off the left end of the graph), the cross-section for this reaction is about 930 bn. Data from National Nuclear Data Center. See also Figs. 1.11 and 3.1

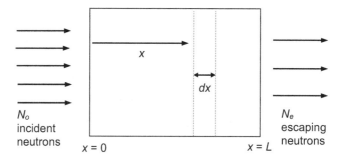

Fig. 2.4 Neutrons penetrating a target of thickness L

$x = L$ will be $P_{react} = 1 - e^{-\sigma nL}$. It follows that if N_o neutrons are incident at the $x = 0$ face, then the number that will be consumed in reactions within the slab will be $N_{react} = N_o\left(1 - e^{-\sigma nL}\right)$. We will use this result in a moment.

Also from (2.7), the number of neutrons that penetrate to distances x and $x + dx$ are given by

$$N_x = N_o e^{-\sigma n x} \tag{2.9}$$

and

$$N_{x+dx} = N_o e^{-\sigma n(x+dx)}. \tag{2.10}$$

Some of the neutrons that reach x will be involved in reactions before reaching $x + dx$, that is, $N_x > N_{x+dx}$. The number of neutrons consumed between x and $x + dx$, designated as dN_x, is then given by

$$dN_x = N_x - N_{x+dx} = N_o e^{-\sigma n x}\left(1 - e^{-\sigma n \, dx}\right). \tag{2.11}$$

If dx is infinitesimal, then $(\sigma \, n \, dx)$ will be very small. This means that we can write $e^{-\sigma n \, (dx)} \sim 1 - \sigma \, n(dx)$, and hence write dN_x as

$$dN_x = N_o e^{-\sigma n x}(\sigma \, n \, dx). \tag{2.12}$$

You should be able to prove that this result is equivalent to differentiating (2.7).

Now, these dN_x neutrons penetrated distance x into the slab before being consumed or diverted in a reaction, so the total travel distance accumulated by all of them in doing so would be $(x \, dN_x)$. The average distance that a neutron will travel before suffering a reaction is given by integrating accumulated travel distances over the length of the slab, and then dividing by the number of neutrons consumed in reactions within the slab, $N_{react} = N_o\left(1 - e^{-\sigma n L}\right)$ from above:

$$\langle x \rangle = \frac{1}{N_{react}} \int_0^L x \, dN_x = \frac{1}{N_o\left(1 - e^{-\sigma n L}\right)} \int_0^L (N_o \sigma n) \, x \, e^{-\sigma n x} dx$$

$$= \frac{1}{\sigma \, n}\left[\frac{1 - e^{-\sigma n L}(1 + \sigma n L)}{1 - e^{-\sigma n L}}\right]. \tag{2.13}$$

If we have a slab of infinite thickness, or, more practically, one such that the product $\sigma n L$ is large, then $e^{-\sigma n L}$ will be small and we will have

$$\langle x \rangle_{(\sigma \, n \, L) \, large} \rightarrow \frac{1}{\sigma \, n}. \tag{2.14}$$

This quantity is known as the *characteristic length* or *mean free path* for the particular reaction quantified by σ. This quantity will figure prominently throughout the remainder of this chapter. If it is computed for an individual cross section such as $\sigma_{fission}$ or $\sigma_{capture}$, one speaks of the mean free path for fission or capture. Such lengths are often designated by the symbol λ with a subscript indicating the type of reaction involved. As an example, consider fission in ^{235}U. The nuclear number density n is 4.794×10^{28} m^{-3}, and the fast-neutron cross section is $\sigma_f = 1.235$ bn $= 1.235 \times 10^{-28}$ m^2, again averaged over the energy spectrum of fission-liberated neutrons. These numbers give $\lambda_f = 16.9$ cm, or about 6.65 inches.

Finally, it should be emphasized that the derivations in this section do not apply to bombarding particles that are *charged*, in which case one has very complex ionization issues to deal with.

2.2 Critical Mass: Bare Core

We now consider critical mass per se. Qualitatively, the concept of critical mass derives from the observation that some species of nuclei fission upon being struck by a bombarding neutron and consequently release secondary neutrons which can potentially go on to induce other fissions, resulting in a chain reaction. However, the development in the preceding section indicates that we can expect that a certain number of neutrons will reach the surface of the mass and escape, particularly if the mass is small. If the density of neutrons within the mass is increasing with time, *criticality* is said to obtain. Whether or not this condition is fulfilled depends on quantities such as the density of the material, its cross-section for fission, the number of neutrons emitted per fission, and the kinetic-energy spectrum of the neutrons. The number of neutrons emitted per fission is designated by the symbol v.

A comment on v is appropriate here. A given fission reaction will release some integer number of neutrons, which on rare occasions could in fact be zero. In carrying out calculations, we will assume an operative *average* number of neutrons per fission. This will inevitably be a decimal number (see Table 2.1), but it should be borne in mind that a more advanced treatment would account for the spectrum of neutron-number emission for a given material when bombarded by neutrons of some spectrum of energies. There is almost no end to the levels of sophistication with which one can approach nuclear-weapons calculations.

To explore the time-dependence of the number of neutrons in a bomb core requires the use of time-dependent *diffusion theory*. In this section we use this theory to

Table 2.1 Threshold Critical Radii and Masses (Untamped; $\alpha = 0$)

Quantity	Unit	^{235}U	^{239}Pu
A	gr mol^{-1}	235.04	239.05
ρ	gr cm^{-3}	18.71	15.6
σ_f	bn	1.235	1.800
σ_{el}	bn	4.566	4.394
v	—	2.637	3.172
n	10^{22} cm^{-3}	4.794	3.930
$\lambda_{fission}$	cm	16.89	14.14
$\lambda_{elastic}$	cm	4.57	5.79
λ_{total}	cm	3.596	4.108
ε	—	1.467	1.090
τ	10^{-9} s	8.635	7.227
d	cm	3.517	2.985
R_O	cm	8.37	6.346
M_O	kg	45.9	16.7

calculate the critical masses of so-called "bare" spherical masses of ^{235}U and ^{239}Pu, the main "active materials" used in fission weapons. The term "bare" is the technical terminology for an *untamped* core. More correctly, we compute critical *radii* which can be transformed into equivalent critical *masses* upon knowing the densities of the materials involved.

The development presented here is based on the derivation in Appendix G of a differential equation which describes the spatiotemporal behavior of the neutron number density N, that is, the number of neutrons per cubic meter within the core. The derivation in Appendix G depends upon on some material developed in Sect. 3.5; it is consequently recommended that both those sections be read in support of this one. Also, be sure not to confuse n and N; the former is the number density of fissile *nuclei*, while the latter is the number density of *neutrons*; both play roles in what follows. Note also that the definition of N here differs from that in the previous section, where it represented a number of neutrons.

Before proceeding, an important limitation of this approach needs to be made clear. Following Serber (1992), I model neutron flow within a bomb core by use of a diffusion equation. A diffusion approach is appropriate if neutron scattering is isotropic. Even if this is not so, a diffusion approach will still be reasonable if neutrons suffer enough scatterings so as to effectively erase non-isotropic angular effects. Unfortunately, neither of these conditions is strictly fulfilled in the case of a uranium or plutonium core: Fast neutrons elastically scattering against uranium show a strong forward-peaked effect. Further, since the mean free path of a fast neutron in ^{235}U, about 3.6 cm, is only about half of the 8.4-cm bare critical radius (see Table 2.1), one cannot help but question the inherent accuracy of the diffusion equation developed in Appendix G. I adopt a diffusion-theory approach for a number of reasons, however. As much of the physics of this area remains classified or at least not easily accessible, we are forced to settle for an approximate model; diffusion theory has the advantage of being analytically tractable at an upper-undergraduate level. In actuality, however, we will see toward the end of this section that the predictions of diffusion theory compare very favorably with *experimentally-measured* critical masses. Also, as shown in Sect. 2.7, a comparison of critical radii as predicted by diffusion theory with those estimated from an openly-published, more exact treatment shows that the two agree to within about 5% for the range of fissility parameters of interest here. We can thus be confident in a diffusion analysis despite its built-in approximations.

Central to any discussion of critical radius are the *fission* and *transport* mean free paths for neutrons, respectively symbolized as λ_f and λ_t. These are given by (2.14) as

$$\lambda_f = \frac{1}{\sigma_f n} \tag{2.15}$$

and

$$\lambda_t = \frac{1}{\sigma_t n}, \tag{2.16}$$

where σ_t is the so-called total or transport cross-section. If neutron scattering is isotropic (which we assume), the transport cross-section is given by the sum of the fission and elastic-scattering cross-sections:

$$\sigma_t = \sigma_f + \sigma_{el}. \tag{2.17}$$

We do not consider here the role of *inelastic* scattering, which affects the situation only indirectly in that it lowers the mean neutron velocity.[1]

For a spherical bomb core, the diffusion theory of Appendix G provides the following differential equation for the time rate of change of the neutron number density:

$$\frac{\partial N}{\partial t} = \frac{v_{neut}}{\lambda_f}(v-1)\,N + \frac{\lambda_t v_{neut}}{3}\left(\nabla^2 N\right), \tag{2.18}$$

where v_{neut} is the average neutron speed and the other symbols are as defined earlier. The first term on the right side of (2.18) corresponds to the growth in the number of neutrons due to fissions, while the second term accounts for neutron loss by their flying out of a volume being considered.

Now, let r represent the usual spherical radial coordinate as measured from the center of the core. Upon assuming a solution for $N(t,r)$ of the form $N(t,r) = N_t(t)\,N_r(r)$, (2.18) can be separated as

$$\frac{1}{N_t}\left(\frac{\partial N_t}{\partial t}\right) = \left(\frac{v-1}{\tau}\right) + \frac{D}{N_r}\left[\frac{1}{r^2}\frac{\partial}{\partial r}\left(r^2\frac{\partial N_r}{\partial r}\right)\right], \tag{2.19}$$

where D is the so-called diffusion coefficient,

---

[1]Eqs. (2.15) and (2.16) assume that the product σnL is large; see the preceding section. For ^{235}U, the values of the square bracket in (2.13) for $L = 10$ cm are 0.267 for $\sigma_{fiss}nL$ and 0.816 for $\sigma_{total}nL$, whereas the large-product approximation assumes that the square bracket will be equal to one. The approximation is more dramatic for the fission mean free path due to its small cross-section. It is thus somewhat surprising that diffusion theory ends up predicting critical masses in close accord with experimentally-measured values; see the discussion following Table 2.1 and Sect. 2.7. As for neglecting inelastic scattering, this is not as drastic as it may seem for a combination of reasons. What matters to the growth of the neutron population is the time τ that a neutron will typically travel before causing another fission; see (2.21). But, if one averages through the many resonance spikes in Fig. 3.1, the fission cross-section for ^{235}U (and ^{239}Pu as well) behaves approximately as $\sigma \sim 1/v_{neut}$, where v_{neut} is the neutron speed. This means that the mean free path for fission is proportional to v_{neut}, which, overall, makes τ independent of v_{neut}. Hence, if a neutron has been either elastically or inelastically scattered, the time for which it will typically travel before causing a subsequent fission is largely independent of its speed. It would then seem that one should add in the inelastic-scattering cross-section when forming the transport cross-section in (2.17). This is true, but another effect comes into play: Elastic scattering is not isotropic. This has the effect of somewhat lowering the effective value of the elastic scattering cross-section. For elements like uranium and plutonium, the two effects largely cancel each other, with the net result that (2.17) is a quite reasonable approximation. Details are given in the Appendix to Serber's *Primer*; see also Soodak (1962), Chap. 3.

$$D = \frac{\lambda_t v_{neut}}{3}, \tag{2.20}$$

and where τ is the mean time that a neutron will travel before causing a fission:

$$\tau = \frac{\lambda_f}{v_{neut}}. \tag{2.21}$$

If the separation constant for (2.19) is defined as α/τ (that is, the constant to which both sides of the equation must be equal), then the solution for the time-dependent part of the neutron density emerges directly as

$$N_t(t) = N_o e^{(\alpha/\tau)t}, \tag{2.22}$$

where N_o represents the neutron density at the center of the core at $t = 0$. N_o would be set by whatever device is used to initiate the chain-reaction. We could have called the separation constant just α, but this form will prove more convenient for subsequent algebra. How α is determined is described following (2.31) below.

Equation (2.22) shows that the time-growth or decay (depending on the sign of α) of the neutron density is exponential. While our main concern for the present is with the *spatial* behavior of N, α will prove to be *very important* throughout this and subsequent sections. We will return to the issue of time-dependence in Sects. 2.5 and 2.6.

With the above definition of the separation constant, the radial part of (2.19) appears as

$$\left(\frac{v-1}{\tau}\right) + \frac{D}{N_r}\left[\frac{1}{r^2}\frac{\partial}{\partial r}\left(r^2\frac{\partial N_r}{\partial r}\right)\right] = \frac{\alpha}{\tau}. \tag{2.23}$$

The first and last terms in (2.23) can be combined; this is why the separation constant was defined as α/τ. On then dividing through by D, we find

$$\frac{1}{d^2} + \frac{1}{N_r}\left[\frac{1}{r^2}\frac{\partial}{\partial r}\left(r^2\frac{\partial N_r}{\partial r}\right)\right] = 0, \tag{2.24}$$

where d is a characteristic length scale,

$$d = \sqrt{\frac{\lambda_f \lambda_t}{3(-\alpha + v - 1)}}. \tag{2.25}$$

Now define a new dimensionless coordinate x according as

$$x = \frac{r}{d}. \tag{2.26}$$

This brings (2.24) to the form

$$\frac{1}{N_r}\left[\frac{1}{x^2}\frac{\partial}{\partial x}\left(x^2\frac{\partial N_r}{\partial x}\right)\right] = -1. \tag{2.27}$$

The solution of this differential equation can easily be verified to be

$$N_r(r) = A\left(\frac{\sin x}{x}\right) + B\left(\frac{\cos x}{x}\right), \tag{2.28}$$

where A and B are constants of integration. There are two terms in the solution because (2.27) is a second-order differential equation. However, both terms need not apply in any given physical situation, and in this case we drop the $(cos\ x)/x$ part of the solution because it would diverge at $x = 0$, that is, at $r = 0$. In Sect. 2.4 we will see a case where both terms are retained.

To determine a critical radius R_C, we need a boundary condition to apply to (2.28). As explained in Appendix G, this takes the form

$$N(R_C) = -\frac{2\lambda_t}{3}\left(\frac{\partial N}{\partial r}\right)_{R_C} = -\frac{2\lambda_t}{3d}\left(\frac{\partial N}{\partial x}\right)_{R_C}. \tag{2.29}$$

On applying this to (2.28), the constant of integration A cancels, and what remains is a transcendental equation for the critical radius:

$$x\cot(x) + \varepsilon x - 1 = 0, \tag{2.30}$$

where

$$\varepsilon = \frac{3d}{2\lambda_t} = \frac{1}{2}\sqrt{\frac{3\lambda_f}{\lambda_t(-\alpha + \nu - 1)}}. \tag{2.31}$$

With fixed values for the density and nuclear constants for some fissile material, Eqs. (2.30) and (2.31) contain two variables: the core radius r (through x) and the exponential factor α (through ε), and the two equations can be solved in two different ways. For both approaches, assume that we are working with material of "normal" density, which we designate as ρ_o. For the first approach, start by looking back to (2.22). If $\alpha = 0$, the neutron number density is neither increasing nor decreasing with time; in this case one has what is called *threshold criticality*. To determine the so-called threshold bare critical radius R_o, set $\alpha = 0$ in (2.25) and (2.31), set the density to ρ_o to determine n, λ_f, and λ_t, solve (2.30) for x, and then get r ($= R_o$) from (2.26). The corresponding threshold bare critical mass M_o then follows from $M_o = (4\pi/3)R_o^3\rho_o$. It is this mass that one usually sees referred to as *the* critical mass; this quantity will figure prominently in the discussion of bomb efficiency in Sects. 2.5 and 2.6.

Table 2.1 shows calculated bare threshold critical radii and masses for ^{235}U and ^{239}Pu. Sources for the fission and elastic-scattering cross-sections appearing in the Table are given in Appendix B; the values quoted therein are used as they are averaged

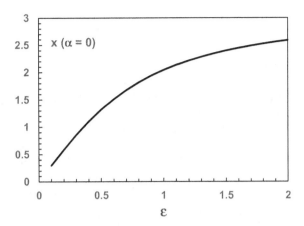

Fig. 2.5 Solution of (2.30) for $\alpha = 0$

over the fission-energy spectra of the two nuclides. The ν values were adopted from the Evaluated Nuclear Data Files (ENDF) maintained by the National Nuclear Data Center at Brookhaven National Laboratory (www.nndc.bnl.gov), and are for neutrons of energy 2 MeV, about the average energy of fission neutrons. The density for ^{235}U is (235/238) times the density of natural uranium, 18.95 gr/cm^3. It is worth noting that the timescales involved in fission-bomb phenomena are remarkably brief: Neutrons travel for only $\tau \sim 1/100$ of a microsecond ($= 10$ ns) between fissions!

Figure 2.5 shows the solution of (2.30) for x when $\alpha = 0$ over the range of ε likely to be of practical interest. Readers should check that the values of ε, d, and radii in Table 2.1 are consistent with this plot. In Sect. 2.7, a similar plot is developed using a parameter akin to ε but of interesting historical provenance.

An important aspect of Fig. 2.5 is that it also applies in cases where the core has been compressed, as in an implosion bomb. For a given core material, ε is independent of density, so the value of x will not change. However, d_{core} is inversely proportional to density, so the critical radius $R_{crit} = xd_{core}$, will be reduced. If the compression factor (the factor by which the density is increased) is C, then R_{crit} will be reduced to R_o/C, and the critical mass, which is proportional to ρR^3, will consequently be reduced by a factor of C^2. For example, consider ^{239}Pu, which has an uncompressed bare critical mass of ~16.7 kg. If $C = 2.5$ [about what was achieved in the *Trinity* test as indicated by analysis of fallout products; Semkow et al. (2006)], the critical mass would reduce to ~2.67 kg. If one were not aware of the bare critical mass but knew that $\varepsilon = 1.09$ for this isotope, Fig. 2.5 indicates $x \sim 2.1$ (the precise value is 2.125). The normal (uncompressed value) of d is 2.99 cm, so the compressed R_{crit} becomes ~(2.1)(2.99/2.5) ~ 2.51 cm. With the density increased to (15.6 g cm^{-3})(2.5) $= 39$ g cm^{-3}, this radius gives a mass of ~2.6 kg, close enough to the precise value. The 6.3 kg core of the *Trinity* device, when compressed, was thus more than critical.

The second way of solving (2.30) begins with assuming that one has a core of some radius $r > R_o$. In this case one will find that (2.30) will be satisfied by some value of $\alpha > 0$, with α increasing as r increases. The rationale here is that since the middle term in (2.30), $\varepsilon x = 3r/2\lambda_t$, is independent of α, we can set r to some desired value; (2.30) can then be solved for x, which gives d from (2.26) and hence α from (2.25). If $\alpha > 0$, the reaction will in principle grow exponentially in time until all of the fissile material is used up, a situation known as "supercriticality" as in the *Trinity* core discussed above.

To see why increasing the radius demands that α must increase, implicitly differentiate (2.30) to show that $d\varepsilon/dx = -\left(1/x\right)^2\left(1 - x^2/\sin^2 x\right)$. This expression demands $d\varepsilon/dx > 0$ for all values of x. From the definition of x, an increase in r (and/or in the density, for that matter) will cause x to increase. To keep (2.30) satisfied means that ε must increase, which, from (2.30), can happen only if α increases.

We come now to a very important point. This is that the condition for threshold criticality can in general be expressed as a constraint on the product $\rho\,r$, where ρ is the mass density of the material and r is the core radius. Recall that ε in (2.30) is independent of the density. Hence, for $\alpha = 0$, (2.30) will be satisfied by some unique value of x which will be characteristic of the material being considered (Fig. 2.5). Since $x = r/d$ and d itself is proportional to $1/\rho$ [see (2.25)], we can equivalently say that the solution of (2.30) demands a unique value of $\rho\,r$ for a given combination of values of σ and ν. If R_o is the bare threshold critical radius for material of normal density ρ_0, then any combination of r and ρ such that $\rho\,r = \rho_0\,R_o$ will also be threshold critical, and any combination such that $\rho\,r > \rho_0\,R_o$ will be supercritical. For a sphere of material of mass M, the mass, density, and radius relate as $M \propto \rho\,r^3$, which means that the "criticality product" $\rho\,r$ can be written as $\rho\,r \propto M/r^2$. This relationship underlies the concept of *implosion* weapons. If a sufficiently strong implosion can be achieved, then one can get away with having less than a "normal" critical mass by starting with a sphere of material of normal density and crushing it to high density by implosion; such weapons inherently make more efficient use of available fissile material than those that depend on a non-implosive mechanism to assemble subcritical components. As described in Sect. 4.2, the implosion technique also helps to overcome predetonation issues with spontaneous fission. The key message here is that there is no *unique* critical mass for a given fissile material.

Lest you think that publishing estimates of critical masses is engaging in revealing classified data, do not be alarmed; such estimates have been available in the public domain for decades. In a review article on fast-neutron reactors, Koch and Paxton (1959) quote a value of 48.7 kg for a spherical assembly of highly enriched uranium (93.9% ^{235}U), and 16.6 kg for a sphere of ^{239}Pu. A 1963 publication of the United States Atomic Energy Commission, "Reactor Physics Constants," a compilation of data for nuclear engineers, lists the *experimentally determined* bare critical mass for 93.9% ^{235}U as 48.8 kg, and that for ^{239}Pu as 16.3 kg. Both values are close to those listed in Table 2.1. Estimating a critical mass is one of the *least* difficult parts of making a nuclear weapon.

Spreadsheet **CriticalityAnalytic.xls** allows users to carry out the above calculations for themselves. This spreadsheet is also used for calculations developed in Sects. 2.3 and 2.5. In its simplest use—corresponding to this section—the user enters five parameters: the density, atomic weight, fission and scattering cross-sections of the core material, and the number of secondary neutrons per fission. The "Goal Seek" function then allows one to solve (2.30) and (2.31) for x (assuming $\alpha = 0$), from which the bare critical radius and mass are computed.

In practice, having available only a single critical mass of fissile material will not produce much of an explosion. The reason for this is that fissioning nuclei give rise to fission products with tremendous kinetic energies. The core consequently very rapidly—within microseconds—heats up and expands, causing its density to drop below that necessary to maintain criticality. In a core comprising only a single critical mass, this will happen at the moment fissions begin, so the chain reaction will quickly fizzle as α falls below zero. To get an explosion of appreciable efficiency, one must start with more than a single critical mass of fissile material, or implode an initially subcritical mass to high density before initiating the explosion. The issue of using more than one critical mass to enhance weapon efficiency is examined in more detail in Sects. 2.5 and 2.6. The effect of using a tamper is examined analytically in Sect. 2.3, and numerically in Sect. 2.6.

To determine the value of the exponential growth factor α for a core of more than one critical mass, it is necessary to solve Eqs. (2.26), (2.30), and (2.31) for α as described following (2.31) above. For the purpose of generating a seed value or simply for making quick estimates, however, an approximate value can be obtained as follows.

Equation (2.28) for the radial dependence of the neutron density appears as

$$N_r(r) = A\left(\frac{\sin x}{x}\right).$$ (2.32)

As a *simplified* boundary condition, assume that $N_r(R_{core}) = 0$, that is, that the neutron density falls to zero at the edge of the core. This is a more restrictive condition that the true boundary condition, (2.29), and will lead to a larger bare threshold critical radius. In this case, (2.32) indicates that we must have $\sin(x) = 0$, or $R/d = \pi$. This will be the case whether a core is supercritical or just threshold critical. If we use subscripts "*core*" and "*o*" to designate a supercritical and bare-threshold-critical core, respectively, then we must have

$$\frac{R_{core}}{d_{core}} = \frac{R_o}{d_o} \Rightarrow \left(\frac{R_o}{R_{core}}\right)^2 = \left(\frac{d_o}{d_{core}}\right)^2.$$ (2.33)

Substitute for d_o and d_{core} from (2.25), setting $\alpha = 0$ in the expression for d_o. The result can then be solved for α_{core}:

$$\alpha_{core} \sim (\nu - 1)\left[1 - \left(R_o/R_{core}\right)^2\right].$$ (2.34)

This result is expressed as an approximation as a reminder that it does not derive from the true boundary condition for neutron diffusion. This simplified boundary condition is explored further in Exercise 2.11.

As an example of how good an estimate (2.34) provides, we consider the Hiroshima *Little Boy* bomb core. It is described in the Preamble that this core had a total mass of ~66 kg. If we take this to be pure ^{235}U of density 18.71 gr/cm^3 for sake of argument, this would correspond to a radius $R_{core} = 9.443$ cm. With $R_o = 8.366$ cm and $\nu = 2.637$ from Table 2.1, (2.34) gives $\alpha_{core} \sim 0.352$. The true value for α for such a core (if pure ^{235}U) is 0.277. The approximation is about 27% high: not terribly accurate, but certainly in the ballpark. (The *Little Boy* core was actually cylindrical, so we have taken some liberty in this example for sake of simplicity.)

To close this section, it is interesting to look briefly at a famous *mis*calculation of critical mass on the part of Werner Heisenberg. At the end of World War II, a number of prominent German physicists including Heisenberg were interned for six months in England and their conversations secretly recorded. This story is detailed in Bernstein (2001); see also Logan (1996) and Bernstein (2002). On the evening of August 6, 1945, the internees were informed that an atomic bomb had been dropped on Hiroshima, and that the energy released was equivalent to about 20,000 tons of TNT. (In actuality, the yield was about 13,000 tons, but this is not the problem with Heisenberg's calculation.) Heisenberg then estimated the critical mass based on this number and a subtly erroneous model of the fission process.

We saw in Sect. 1.6 that complete fission of 1 kg of ^{235}U liberates energy equivalent to about 17 kilotons of TNT. Heisenberg predicated his estimate of the critical mass on assuming that about 1 kg of material did in fact fission. One kilogram of ^{235}U corresponds to about $\Omega \sim 2.56 \times 10^{24}$ nuclei. Assuming that on average $\nu = 2$ neutrons are liberated per fission, then the number of generations G necessary to fission the entire kilogram would be $\nu^G = \Omega$. Solving for G gives $G = \ln(\Omega)/\ln(\nu) \sim 81$, which Heisenberg rounded to 80. So far, this calculation is fine. He then argued that as neutrons fly around in the bomb core, they will randomly bounce between nuclei, traveling a mean distance λ_f before causing fissions; λ_f is the mean free path between fissions as in (2.15) above. From Table 2.1, $\lambda_f \sim 17$ cm for ^{235}U, but, at the time, Heisenberg took $\lambda_f \sim 6$ cm. Since a random walk of G steps where each is of length λ_f will take one a distance $r \sim \lambda_f \sqrt{G}$ from the starting point, he estimated a critical radius of $r \sim (6 \text{ cm})\sqrt{80} \sim 54$ cm. This would correspond to a mass of some 12,500 kg, roughly 13 tons! Given that only one kilogram of uranium fissioned, this would be a fantastically inefficient weapon. Such a bomb and its associated tamper, casing, and instrumentation would represent an unbearably heavy load for a World War II-era bomber.

The problem with Heisenberg's calculation was that he imagined the fission process to be created by a single neutron that randomly bounced throughout the bomb core, begetting secondary neutrons along the way. Further, his model is too stringent; there is no need for every neutron to cause a fission; many neutrons escape. In the days following August 6, Heisenberg revised his model, arriving at the diffusion theory approach described in this section.

2.3 Critical Mass: Tamped Core

In the preceding section, it was shown how to calculate the critical mass of a "bare" sphere of fissile material. In this section we develop a model to account for the presence of a tamper. The discussion here draws from the preceding section and from Serber (1992), Bernstein (2002), and especially Reed (2009).

The idea behind a tamper is to surround the fissile core with a shell of dense material, as suggested in Fig. 2.6. This serves two purposes: (i) It reduces the critical mass, and (ii) It slows the inevitable expansion of the core, allowing more time for fissions to occur until the core density drops to the point where criticality no longer holds. The reduction in critical mass occurs because the tamper will reflect some escaped neutrons back into the core; indeed, the modern name for a tamper is "reflector," but I retain the historical terminology here. This effect is explored in this section. Estimating the distance over which an *untamped* core expands before criticality no longer holds is analyzed in Sect. 2.5. This slowing effect is difficult to model analytically, but can be treated approximately with a numerical model; this is done in Sect. 2.6; a rough analytic model is presented in Sect. 2.8.1.

The discussion here parallels that in Sect. 2.2. Neutrons that escape from the core will diffuse into the tamper. If the tamper material is not fissile, we can describe the behavior of neutrons within it via (2.18) without the neutron-production term, that is, without the first term on the right side:

$$\frac{\partial N_{tamp}}{\partial t} = \frac{\lambda_{trans}^{tamp} v_{neut}}{3} \left(\nabla^2 N_{tamp} \right), \tag{2.35}$$

where N_{tamp} is the number density of neutrons within the tamper and λ_{trans}^{tamp} is their transport mean free path. v_{neut} is the average neutron speed within the tamper, which we will later assume for sake of simplicity to be the same as that within the core. We are assuming that the tamper does not capture neutrons; otherwise, we would have to add a term to (2.35) to represent that effect.

Fig. 2.6 Schematic illustration of a tamped bomb core

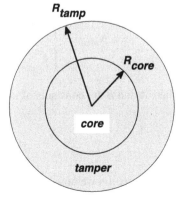

Superscripts and subscripts *tamp* and *core* will be used liberally here as it will be necessary to join *tamper* physics to *core* physics via suitable boundary conditions. As was done in Sect. 2.2, take a trial solution for N_{tamp} of the form $N_{tamp}(t, r) = N_t^{tamp}(t) N_r^{tamp}(r)$, where $N_t^{tamp}(t)$ and $N_r^{tamp}(r)$ are respectively the time-and space dependences of N_{tamp}; r is the usual spherical radial coordinate measured from the center of the core. Upon substituting this into (2.35) we find, in analogy to (2.19),

$$\frac{1}{N_t^{tamp}}\left(\frac{\partial N_t^{tamp}}{\partial t}\right) = \left(\frac{\lambda_{trans}^{tamp} v_{neut}}{3}\right)\frac{1}{N_r^{tamp}}\left[\frac{1}{r^2}\frac{\partial}{\partial r}\left(r^2\frac{\partial N_r^{tamp}}{\partial r}\right)\right]. \qquad (2.36)$$

Define the separation constant here to be δ/τ, where τ is the mean time that a neutron will travel *in the core* before causing a fission, that is, as defined in (2.21):

$$\tau = \frac{\lambda_{fiss}^{core}}{v_{neut}}. \qquad (2.37)$$

While it may seem strange to invoke a *core* quantity when dealing with diffusion in the *tamper*, this choice is advantageous in that the neutron velocity v_{neut}, which we assume to be the same in both materials, will cancel out in later algebra. This choice is *not* equivalent to assuming at the outset that the core and tamper separation constants are the same, as δ may be different from the exponential factor α of Sect. 2.2. However, we will find that boundary conditions demand that they too must be equal. This choice of separation constant renders (2.36) as

$$\frac{1}{N_t^{tamp}}\left(\frac{\partial N_t^{tamp}}{\partial t}\right) = \left(\frac{\lambda_{trans}^{tamp} v_{neut}}{3}\right)\frac{1}{N_r^{tamp}}\left[\frac{1}{r^2}\frac{\partial}{\partial r}\left(r^2\frac{\partial N_r^{tamp}}{\partial r}\right)\right] = \frac{\delta}{\tau}. \qquad (2.38)$$

The solution of (2.38) depends on whether δ is positive, negative, or zero; the latter choice corresponds to threshold criticality in analogy to $\alpha = 0$ in Sect. 2.2. The situations of practical interest will be $\delta \geq 0$, in which case the solutions have the form

$$N_{tamp} = \begin{cases} \dfrac{A}{r} + B & (\delta = 0) \\ e^{(\delta/\tau)t}\left\{A\dfrac{e^{r/d_{tamp}}}{r} + B\dfrac{e^{-r/d_{tamp}}}{r}\right\} & (\delta > 0), \end{cases} \qquad (2.39)$$

where A and B are constants of integration (different for the two cases), and where

$$d_{tamp} = \sqrt{\frac{\lambda_{trans}^{tamp}\lambda_{fiss}^{core}}{3\delta}}. \qquad (2.40)$$

The situation we now have is that the neutron density in the core is described by (2.22) and (2.28) as

$$N_{core} = A_{core} e^{(\alpha/\tau)t} \frac{\sin(r/d_{core})}{r}, \tag{2.41}$$

where A_{core} is different from the A in (2.39) and d_{core} given by (2.25):

$$d_{core} = \sqrt{\frac{\lambda_{fiss}^{core} \lambda_{trans}^{core}}{3(-\alpha + \nu - 1)}}, \tag{2.42}$$

while the neutron density in the tamper is given by (2.39) and (2.40).

The question at this point is: "What boundary conditions apply in order that we have a physically reasonable solution?" Let the core have radius R_{core}, and let the outer radius of the tamper be R_{tamp}; we assume that the inner edge of the tamper is snug against the core. First consider the core/tamper interface. If no neutrons are created or lost at this interface, then it follows that both the density and flux of neutrons across the interface must be continuous. That is, we must have

$$N_{core}(R_{core}) = N_{tamp}(R_{core}), \tag{2.43}$$

and, from (6.97) of Appendix G,

$$\lambda_{trans}^{core} \left(\frac{\partial N_{core}}{\partial r} \right)_{R_{core}} = \lambda_{trans}^{tamp} \left(\frac{\partial N_{tamp}}{\partial r} \right)_{R_{core}}. \tag{2.44}$$

Equation (2.44) accounts for the effect of any neutron reflectivity of the tamper via λ_{trans}^{tamp}. In writing (2.44), we have assumed that the speed of neutrons within the core and tamper is the same, and hence cancels.

In addition, we must consider what is happening at the outer edge of the tamper. If there is no "backflow" of neutrons from the outside, then the situation is analogous to the boundary condition of (2.29) that was applied to the outer edge of the untamped core:

$$N_{tamp}(R_{tamp}) = -\frac{2}{3} \lambda_{trans}^{tamp} \left(\frac{\partial N_{tamp}}{\partial r} \right)_{R_{tamp}}. \tag{2.45}$$

Applying (2.43)–(2.45) to (2.39)–(2.42) results, after some algebra, in the following equations of constraint:

$$\left[1 + \frac{2R_{thresh}\lambda_{trans}^{tamp}}{3R_{tamp}^2} - \frac{R_{thresh}}{R_{tamp}} \right] \left[\left(\frac{R_{thresh}}{d_{core}} \right) \cot \left(\frac{R_{thresh}}{d_{core}} \right) - 1 \right]$$

$$+ \frac{\lambda_{trans}^{tamp}}{\lambda_{trans}^{core}} = 0, \quad (\delta = 0) \tag{2.46}$$

and, for $\delta > 0$,

$$e^{2(x_{ct}-x_t)} \left[\frac{x_c \cot x_c - 1 - \lambda (x_{ct} - 1)}{R_{tamp} + 2\lambda_{trans}^{tamp}(x_t - 1)/3} \right] = \left[\frac{x_c \cot x_c - 1 + \lambda (x_{ct} + 1)}{R_{tamp} - 2\lambda_{trans}^{tamp}(x_t + 1)/3} \right],$$
(2.47)

where

$$\left. \begin{aligned} x_{ct} &= R_{core} / d_{tamp} \\ x_c &= R_{core} / d_{core} \\ x_t &= R_{tamp} / d_{tamp} \\ \lambda &= \lambda_{trans}^{tamp} / \lambda_{trans}^{core} \end{aligned} \right\}.$$
(2.48)

It is also necessary to demand that $\alpha = \delta$, as otherwise the fact that (2.43)–(2.45) must also hold as a function of *time* would be violated. Some comments on these results follow. Comments on (2.46) will be particularly extensive.

Equation (2.46) corresponds to *tamped threshold criticality*, where $\alpha = \delta = 0$. Once values for the d's and λ's are given, there are two ways to use this expression. First, if a core mass which is bare-threshold *sub*-critical is specified, use its radius as R_{thresh} and solve (2.46) for R_{tamp}, the tamper outer radius which will just render the core critical. The tamper mass can then be determined from the two radii. Figure 2.7 shows the mass of various tamper materials necessary to render just critical a range of uncompressed ^{235}U core masses; the geometry is assumed to be spherical [DU = depleted uranium, essentially pure ^{238}U; Al = aluminum; BeO = beryllium oxide, WC = tungsten-carbide (steel)]. The curves converge at the untamped critical mass for ^{235}U, 45.9 kg (Table 2.1). In the case of the Hiroshima *Little Boy* bomb, which

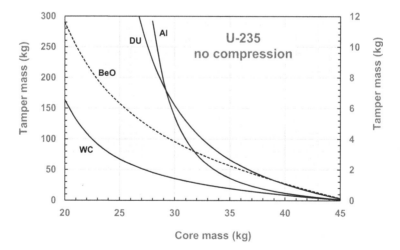

Fig. 2.7 Mass of snugly-fitting tampers which will just render threshold critical a given core mass of uncompressed pure ^{235}U. The dashed curve for BeO is to be read on the right axis; all others are read on the left axis

used a tungsten-carbide tamper, extending this analysis to lower masses shows that a core of 16.17 kg would be rendered critical by such a tamper with a mass of 552 kg (beyond the range of the left side of Fig. 2.7). This mass is close to the actual *Little Boy* tamper mass (see the Preamble), so we can conclude that *Little Boy* utilized about (53/16.17) ~ 3 tamped threshold critical masses of ^{235}U. Table 2.2 lists values of λ_{trans}^{tamp} for some common materials that might be used as tampers. Beryllium oxide is a very desirable tamper material on account of its very low neutron-capture cross-section. In computing these mean-free-paths for materials with more than one isotope, effective cross-sections need to be computed from weighted averages; also, if a material is molecular (e.g., BeO or WC), the cross-sections of all contributing elements have to be added. Figure 2.8 shows an equivalent plot for uncompressed ^{239}Pu.

Table 2.2 Adopted parameters for common tamper materials. Lead and tungsten both have several naturally-occurring isotopes; values given here are abundance-weighted averages

Material	A gr mol^{-1}	ρ gr cm^{-3}	σ_{el} bn	λ_{trans}^{tamp} cm	Infinite tamper critical mass (kg)	
					^{235}U	^{239}Pu
Al	26.982	2.699	2.967	5.595	21.9	9.6
BeO	25.01	3.02	5.412	2.541	8.9	3.9
DU (^{238}U)	238.05	18.95	4.804	4.342	16.4	7.2
WC	195.85	15.63	6.587	3.159	11.4	5.0

DU = Depleted Uranium
WC = tungsten carbide

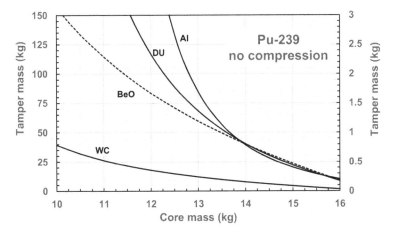

Fig. 2.8 Mass of snugly-fitting tampers which will just render threshold critical a given core mass of uncompressed pure ^{239}Pu. The dashed curve for BeO is to be read on the right axis; all others are read on the left axis

The second way of using (2.46) is to specify R_{tamp}, and then solve for R_{thresh}, the radius of a core which would just be critical for the specified tamper outer radius. This can be a handy calculation if the size of your bomb is limited in advance by some condition such as the diameter of a missile tube.

Figures 2.9 and 2.10 illustrate a very general approach to solving (2.46). This is based on first recasting it in the form

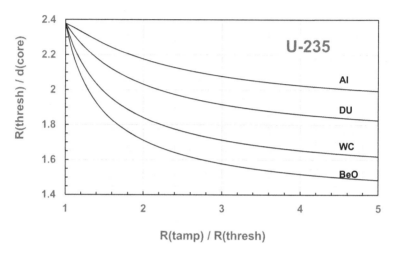

Fig. 2.9 R_{thresh}/d_{core} versus R_{tamp}/R_{thresh} for four commonly-used tampers in combination with ^{235}U ($\varepsilon = 1.467$). The curves converge at $R_{thresh}/d_{core} = 2.379$, corresponding to a bare core

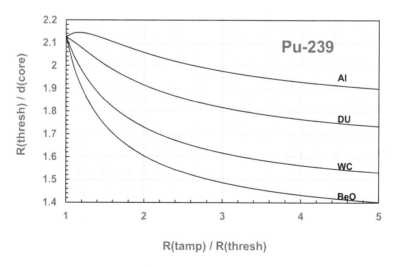

Fig. 2.10 R_{thresh}/d_{core} versus R_{tamp}/R_{thresh} for four commonly-used tampers in combination with ^{239}Pu ($\varepsilon = 1.090$). The curves converge at $R_{thresh}/d_{core} = 2.125$, corresponding to a bare core

$$\left[1 + \left(\frac{\lambda}{\varepsilon}\right)\left(\frac{d_{core}}{R_{thresh}}\right)\left(\frac{R_{thresh}}{R_{tamp}}\right)^2 - \left(\frac{R_{thresh}}{R_{tamp}}\right)\right]\left[\left(\frac{R_{thresh}}{d_{core}}\right)\cot\left(\frac{R_{thresh}}{d_{core}}\right) - 1\right] + \lambda = 0,$$
(2.49)

where ε is as defined in (2.31) and where

$$\lambda = \frac{\lambda_{trans}^{tamp}}{\lambda_{trans}^{core}}.$$
(2.50)

Once the core material is chosen, the value of ε is fixed; values for ^{235}U and ^{239}Pu are given in Table 2.1. For chosen values of ε and λ, Eq. (2.49) can be solved numerically for R_{thresh}/d_{core} for a given value of R_{tamp}/R_{thresh}; this latter quantity must by definition be ≥ 1. A graph of R_{thresh}/d_{core} versus R_{tamp}/R_{thresh} can then be built up. Figures 2.9 and 2.10 show such graphs for the same tamper materials as Figs. 2.7 and 2.8 for ^{235}U and ^{239}Pu, respectively. In Fig. 2.9, the curves converge to $R_{thresh}/d_{core} = 2.379$ at $R_{tamp}/R_{thresh} = 1$, which corresponds to a bare ^{235}U core; similarly for ^{239}Pu in Fig. 2.10, where they converge at $R_{thresh}/d_{core} = 2.125$.

The *Little Boy* example above can be reconstructed from Fig. 2.9, except that the question is now posed in reverse to ask: What tamper mass will be needed to render a core of mass 16.17 kg just critical? A ^{235}U core of mass 16.17 kg will have a radius R_{thresh} of ~5.91 cm, or, from the value of d_{core} in Table 2.1, R_{thresh}/d_{core} ~ 5.91/3.52 ~ 1.68. Eyeballing across the graph shows that a horizontal line at R_{thresh}/d_{core} ~ 1.68 will cross the curve for WC at R_{tamp}/R_{thresh} ~ 3.5. This corresponds to R_{thresh} ~ 20.7 cm, which would give a tamper thickness of ~14.8 cm and a mass of ~ 570 kg, close to that described above. Readers are cautioned that small changes in input parameters can have significant effects on results.

A converse approach to using Figs. 2.9 and 2.10 would be to decide on a value of R_{tamp}/R_{thresh} in advance, and then determine R_{core}/d for a given value of λ. There are various ways to use all of these plots.

An important aspect of Figs. 2.9 and 2.10 is that, like Fig. 2.5, they will also work for cases where the core and tamper have been compressed; this is not the case for Figs. 2.7 and 2.8. If the density of both materials is increased by the same factor, λ will not be affected. (If the density changes are different, which is likely to be the case because materials have different bulk moduli, it is easy to compute the new λ: Individual λ's are simply inversely proportional to their densities). As with Fig. 2.5, the only factor that will be different will be d_{core}.

The curve for aluminum in Fig. 2.10 shows a slight maximum for small values of R_{tamp}/R_{thresh}, which would indicate a value of R_{thresh}/d_{core} *greater* than that for a bare core, a clearly nonsensical result. The reason for this is that such a core would be very thin in comparison to its transport mean free path, for which a diffusion analysis is not accurate.

Here is a hypothetical exercise using Fig. 2.10 and Tables 2.1 and 2.2. Suppose that terrorists have stolen 12 kg of ^{239}Pu. They also have 150 kg of Aluminum with which they can fashion a tamper. They have no implosion technology, so both materials are at normal density. Can they make a threshold-critical device, assuming spherical geometry? From Fig. 2.8, it can be seen directly that the answer is no. To approach this with Fig. 2.10, we determine what tamper outer radius is necessary to achieve criticality, and then see if 150 kg of Al is sufficient to meet that radius. 12 kg of ^{239}Pu would form a sphere of radius 5.68 cm. With $d_{core} = 2.985$ cm, we are asking for $R_{thresh}/d_{core} = 5.68/2.99 = 1.90$. Reading horizontally across Fig. 2.10 at $R_{thresh}/d_{core} = 1.90$ shows that $R_{tamp}/R_{thresh} \sim 5$, or $R_{tamp} \sim 28.4$ cm. This corresponds to a tamper mass of ~ 250 kg, so the answer is again no. (More precisely, one needs ~ 230 kg of Al to just achieve criticality for such a core.)

One further way of plotting Eq. (2.46) is shown in Fig. 2.11: creating curves of threshold-critical core/tamper mass combinations for given compression ratios. The limitation of a plot like this is that it becomes too cluttered if one attempts to display data for more than one tamper material at once.

As for using Eqs. (2.47) and (2.48): Refer to the second way of solving (2.46) above, where R_{thresh} is determined for a given value of R_{tamp}. Keep R_{tamp} fixed to its chosen value. Now choose a core radius $R_{core} > R_{thresh}$ to use in (2.47) and (2.48). Then solve (2.47) numerically for δ $(= \alpha)$, which enters the d's and x's of (2.47) and (2.48) through (2.40) and (2.42). The value of knowing α will become clear when the efficiency and yield calculations of Sects. 2.5 and 2.6 are developed.

A special-case application of (2.46) can be used to get a sense of how dramatically the presence of a tamper decreases the threshold critical mass. Suppose that the tamper is very thick, $R_{tamp} \gg R_{thresh}$. In this case (2.46) reduces to

$$\left(R_{thresh}/d_{core}\right) \cot \left(R_{thresh}/d_{core}\right) = 1 - \left(\lambda_{trans}^{tamp}/\lambda_{trans}^{core}\right). \qquad (2.51)$$

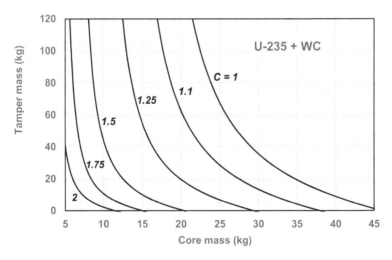

Fig. 2.11 Core/tamper mass combinations that are threshold critical for various compression ratios ($C = 1$, 1.1, 1.25, 1.5, 1.75, 2) for a ^{235}U core and WC tamper

Now consider two sub-cases. The first is that the tamper is in fact a vacuum. Since empty space would have essentially zero cross-section for neutron scattering, this is equivalent to specifying $\lambda_{trans}^{tamp} = \infty$, in which case (2.51) becomes

$$\left(R_{thresh} / d_{core} \right) \cot \left(R_{thresh} / d_{core} \right) = -\infty. \tag{2.52}$$

This can only be satisfied if

$$\left(\frac{R_{thresh}}{d_{core}} \right)_{vacuum\,tamper} = \pi. \tag{2.53}$$

This result is equivalent to assuming a simplified boundary condition of no neutron loss at the surface of an *untamped* core; see Sect. 2.8.1; you might want to reflect on why this would be. The second sub-case is more realistic in that we imagine a thick tamper with a non-zero transport mean free path. For simplicity, assume that $\lambda_{trans}^{core} \sim \lambda_{trans}^{tamp}$, that is, that the neutron-scattering properties of the tamper are much like those of the core. In this case, (2.51) becomes

$$\left(R_{thresh} / d_{core} \right) \cot \left(R_{thresh} / d_{core} \right) = 0. \tag{2.54}$$

The solution here is

$$\left(\frac{R_{thresh}}{d_{core}} \right)_{\substack{thick\,tamper \\ finite\,cross-scetion}} = \frac{\pi}{2}, \tag{2.55}$$

exactly one-half the value of the vacuum-tamper case. To summarize: With an infinitely-thick tamper of finite transport mean free path, the threshold critical radius is one-half of what it would be if no tamper were present at all. A factor of two in radius means a factor of eight in mass, so the advantage of using a tamper is dramatic, even aside from the issue of any retardation of core expansion. This factor of two in critical radius is predicated on an unrealistic assumption for the tamper thickness and so we cannot expect such a dramatic effect in reality, but Figs. 2.7 and 2.8 indicate that the effects are dramatic enough.

For any specific infinitely-thick tamper, cast (2.51) in the form

$$\left(R_{thresh} / d_{core} \right) \cot \left(R_{thresh} / d_{core} \right) - 1 + \lambda = 0. \tag{2.56}$$

This can be solved for (R_{thresh}/d_{core}) as a function of λ; the result is shown in Fig. 2.12.

Purely empirically, the curve in Fig. 2.12 can be expressed as

$$\left(R_{thresh} / d_{core} \right)_{\substack{infinite \\ tamper}} \sim 1.56\,\lambda^{0.38} \sim \frac{\pi}{2}\,\lambda^{0.38}. \tag{2.57}$$

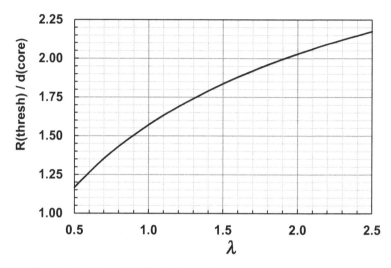

Fig. 2.12 Numerical solution of (2.56) for (R_{thresh}/d_{core}) for infinitely thick tampers

Further details on a historical connection of this expression can be found in Reed (2015). Minimal critical masses for various tampers for ^{235}U and ^{239}Pu are listed in the last two columns of Table 2.2. The smallness of some of these values testifies to why non-proliferation agencies are very concerned with seemingly small amounts of fissile materials.

A comment on beryllium-oxide, which Figs. 2.7 and 2.8 indicate makes for an excellent tamper material. BeO was not used during the Manhattan Project because beryllium was than a fairly rare element; also, if inhaled, it can be toxic.

2.4   Critical Mass: Tamped Composite Core

This section takes up the concept of a tamped *composite* bomb core. A composite core is one where an inner sphere of fissile material is surrounded by an outer sphere of a different fissile material, with both enclosed in a tamper. This is illustrated in Fig. 2.13, where the inner and outer cores have radii R_1 and R_2, and the tamper has outer radius R_{tamp}.

Composite cores were considered during the Manhattan Project, but not utilized in order to stick with proven designs following the *Trinity* test and the successful *Little Boy* drop at Hiroshima. The motivation for considering such cores, however, was that the rate of bomb production could be maximized by combining ^{235}U and ^{239}Pu with masses dictated by the physics of criticality and the ratio of their production rates. An additional advantage of a composite core is that it would involve a much reduced spontaneous-fission background compared with a ^{239}Pu-only core because of the much smaller amount of that material involved; see Sect. 4.2. Had composite

Fig. 2.13 Conceptual sketch of composite core with tamper. The inner and outer cores are presumed to be made of fissile material; the tamper is non-fissile. This sketch is purely schematic

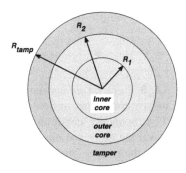

cores been used at that time, the more efficient implosion-based *Fat Man* design would have been employed, and this is what is assumed here. The first composite-core weapon was tested in Operation Sandstone in 1948, and a detailed study of the World War II situation can be found in Reed (2020a). The key physics question of interest here is: How does the critical mass of a composite core bomb, under, say, a compression ratio of *Trinity's* 2.5, compare to that for a ^{239}Pu-only bomb?

The physics of this situation can be analyzed with the solutions of the diffusion equation and the boundary conditions developed in the preceding two sections. The analysis is restricted to threshold criticality, that is, where the exponential growth parameter $\alpha = 0$.

For the inner core (subscript "1"), the neutron density $N_1(r)$ is given by (2.28) without the $r = 0$ divergent cosine term:

$$N_1(r) = A \frac{d_1}{r} \sin\left(\frac{r}{d_1}\right), \tag{2.58}$$

where d_1 is the characteristic length scale of (2.25), with values appropriate to this material:

$$d_1 = \sqrt{\frac{\lambda_1^{fiss} \lambda_1^{tran}}{3(\nu_1 - 1)}}. \tag{2.59}$$

For the outer core (subscript "2"), both terms in (2.28) are retained:

$$N_2(r) = B \frac{d_2}{r} \sin\left(\frac{r}{d_2}\right) + C \frac{d_2}{r} \cos\left(\frac{r}{d_2}\right), \tag{2.60}$$

with

$$d_2 = \sqrt{\frac{\lambda_2^{fiss} \lambda_2^{tran}}{3(\nu_2 - 1)}}. \tag{2.61}$$

For the tamper, the threshold solution of the diffusion equation is (2.39),

$$N_{tamp}(r) = \frac{D}{r} + E. \tag{2.62}$$

(A, B, C, D, E) are constants of integration to be eliminated upon application of boundary conditions, as described in what follows.

Five boundary conditions dictate the continuity of neutron density and flux between the inner and outer cores, the outer core and the tamper, and at the outer edge of the tamper. These are, from (2.43)–(2.45),

$$N_1(R_1) = N_2(R_1), \tag{2.63}$$

$$\lambda_1^{tran}\left(\frac{\partial N_1}{\partial r}\right)_{R_1} = \lambda_2^{tran}\left(\frac{\partial N_2}{\partial r}\right)_{R_1}, \tag{2.64}$$

$$N_2(R_2) = N_{tamp}(R_2), \tag{2.65}$$

$$\lambda_2^{tran}\left(\frac{\partial N_2}{\partial r}\right)_{R_2} = \lambda_{tamp}^{tran}\left(\frac{\partial N_{tamp}}{\partial r}\right)_{R_2}, \tag{2.66}$$

and

$$N_{tamp}(R_{tamp}) = -\frac{2}{3}\lambda_{tamp}^{tran}\left(\frac{\partial N_{tamp}}{\partial r}\right)_{R_{tamp}}. \tag{2.67}$$

The algebra to eliminate A, B, C, D, and E is lengthy; the result is an expression which involves, aside from the usual nuclear parameters, only the three radii (R_1, R_2, R_{tamp}). A reader who wishes to test his or her algebra skills can verify that this expression can be written as

$$U - QV = 0. \tag{2.68}$$

U, Q, and V are

$$U = \left(\frac{B}{C}\right)\frac{\sin x_3}{x_3} + \frac{\cos x_3}{x_3}, \tag{2.69}$$

$$Q = \psi\left\{\left(\frac{B}{C}\right)[-\sin x_3 + x_3\cos x_3] - [\cos x_3 + x_3\sin x_3]\right\}, \tag{2.70}$$

and

$$V = \frac{2\lambda_{tamp}^{tran}}{3R_{tamp}^2} - \frac{1}{R_{tamp}} + \frac{1}{R_2}. \tag{2.71}$$

The various quantities appearing in these expressions are

$$(x_1, \ x_2, \ x_3) \ = \ \left(\frac{R_1}{d_1}, \ \frac{R_1}{d_2}, \ \frac{R_2}{d_2} \right), \tag{2.72}$$

$$\left(\frac{B}{C} \right) \ = \ \frac{[\delta \, Y \sin x_1 \ + \ Z \cos x_2]}{[\delta \, W \sin x_1 \ - \ Z \sin x_2]}, \tag{2.73}$$

and

$$\psi \ = \ -\frac{\lambda_2^{tran} d_2}{\lambda_{tamp}^{tran}}, \tag{2.74}$$

with

$$W \ = \ [-\sin x_2 \ + \ x_2 \cos x_2], \tag{2.75}$$

$$Y \ = \ [\cos x_2 \ + \ x_2 \sin x_2], \tag{2.76}$$

$$Z \ = \ \lambda[-\sin x_1 \ + \ x_1 \cos x_1], \tag{2.77}$$

where

$$\lambda \ = \ \frac{\lambda_1^{tran}}{\lambda_2^{tran}} \frac{d_1}{d_2} \ = \ \frac{\lambda_1^{tran}}{\lambda_2^{tran}} \delta. \tag{2.78}$$

For δ in (2.73), see (2.78). Despite the messiness of this solution, only one unknown can be solved for, because the entire problem reduces to satisfying (2.68). For example, if R_1 and R_{tamp} are specified, then one can solve numerically for R_2. The masses and radii are then related by

$$R_1 \ = \ \left(\frac{3M_1}{4\pi \, \rho_1} \right)^{1/3}, \tag{2.79}$$

$$R_2 \ = \ \left[\frac{3}{4\pi} \left(\frac{M_1}{\rho_1} \ + \ \frac{M_2}{\rho_2} \right) \right]^{1/3}, \tag{2.80}$$

and

$$R_{tamp} \ = \ \left[\frac{3}{4\pi} \left(\frac{M_1}{\rho_1} \ + \ \frac{M_2}{\rho_2} \ + \ \frac{M_{tamp}}{\rho_{tamp}} \right) \right]^{1/3}. \tag{2.81}$$

Any compression involved can be accounted for by multiplying the densities by the desired compression ratio.

Figure 2.14 shows the results of such calculations, assuming a 230 kg DU tamper and compression ratio 2.5 (this was approximately the total tamper mass of *Fat Man*—see the Preamble). Spreadsheets **DualCore(Pu + U).xlsm** and **DualCore(U + Pu).xlsm** were used for these computations; the first assumes a ^{239}Pu inner core with a ^{235}U outer core, and the second the reverse. For such a tamper, the critical mass for ^{235}U alone is 3.5 kg, and that for ^{239}Pu is 1.44 kg; this is why the solid curve (^{235}U inner core) indicates that no outer core is necessary when the mass of ^{235}U is 3.5 kg, and similarly why the dashed curve (^{239}Pu inner core) falls to zero at 1.44 kg. As an example of using this graph, suppose that you have 2 kg of ^{235}U available to use as the inner core. You would then need ~ 0.75 kg of ^{239}Pu for the outer core. The total mass would be less than that for a ^{235}U core alone; this is because of the greater fissility of ^{239}Pu. On the other hand, if you have only 0.7 kg of ^{239}Pu available to use as the inner core, you will need ~1.4 kg of ^{235}U to achieve criticality, for a total of 2.1 kg as opposed to 1.44 kg for a ^{239}Pu-only bomb. But if you have this 0.7 kg of ^{239}Pu available, is it perhaps more efficient to use it as the outer core instead? No: you would need an inner core of ~2 kg of ^{235}U to achieve criticality, or a total of 2.7 kg. It is better to put the more-fissile material as the inner core. The total mass of core material actually used is another issue, dictated by considerations of desired yield and the space available within the implosion assembly.

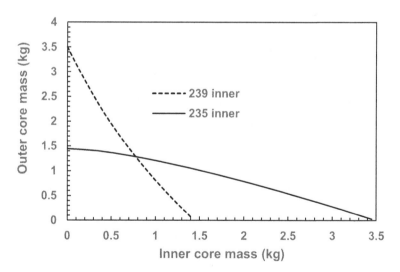

Fig. 2.14 Outer core mass necessary to render critical a given inner core mass for a 230 kg DU tamper and compression ratio 2.5

2.5 Estimating Yield—Analytic

Material in this section is adopted from Reed (2007).

In the preceding sections, we examined how to estimate critical masses for bare, tamped, and tamped composite cores of fissile material. The analysis in Sect. 2.2 revealed that the threshold bare critical mass of ^{235}U is about 46 kg. In Sect. 1.6, we saw that complete fission of 1 kg of ^{235}U liberates energy equivalent to that of about 17 kilotons of TNT. Given that the *Little Boy* uranium bomb that was dropped on Hiroshima used about 53 kg of ^{235}U and is estimated to have had an explosive yield of only about 13 kilotons, we can infer that it must have been rather inefficient. The purpose of this section is to explore what factors dictate the efficiency of a fission weapon and to show how one can estimate that efficiency.

This section is the first of several in this Chapter and in Chap. 4 devoted to the questions of weapon efficiency and yield. In this section, these issues are examined purely analytically. The advantage of an analytic approach is that it is helpful for establishing a sense of how the efficiency depends on the parameters involved: The mass and density of the core and the values of various nuclear constants. However, conditions inside an exploding bomb core evolve very rapidly as a function of time, and this evolution cannot be fully captured with analytic approximations. To get a sense of the time-evolution of the process, it is necessary to numerically integrate the core conditions as a function of time, tracking core size, expansion rate, pressure, neutron density, and energy release along the way. Such an analysis is the subject of the next section; these two sections therefore closely complement each other and should be read as a unit. Bomb efficiency and yield can also be affected by various phenomena that can trigger the chain-reaction before the weapon core has reached its fully assembled state; these issues are explored in Chap. 4.

In the present section we consider only *untamped* cores for sake of simplicity; a tamped core is simulated numerically in Sect. 2.6.

To begin, it is helpful to appreciate that the efficiency of a nuclear weapon involves three distinct time scales. The first is mechanical in nature: The time required to assemble the subcritical fissile components into a critical assembly before fission is initiated. In principle, this time can be as long as is desired, but in practice it is constrained by the occurrence of spontaneous fissions, which could lead to reaction-triggering stray neutrons during the assembly period.

What is the order of magnitude of the assembly time? This was discussed in the Preamble, where it was shown that for a "gun-type" bomb where a "projectile" piece of fissile material is fired as a shell inside an artillery barrel toward a mating "target" piece of fissile material, about 200 microseconds will be required to achieve full core assembly from the moment when the leading edge of the projectile meets the target piece. As we will see in Sect. 4.2, the rate of spontaneous fission was not an issue for a uranium core over this time, but was such a problem for plutonium as to necessitate development of the implosion mechanism for triggering those weapons. So far as the present section is concerned, however, the essential idea is that if the spontaneous fission probability can be kept negligible during the assembly time (which we assume), the efficiency of the weapon is dictated by the two other time scales.

The first of these other time scales is nuclear in nature. Once fission has been initiated, some time will be required for all of the fissile material to be consumed. This time we call t_{fiss}. The other time scale is again mechanical. As soon as fissions have been initiated, the core will begin to expand due to the extreme gas pressure of the fission fragments. This expansion will lead after a time t_{crit} to loss of criticality, after which the reaction rate will diminish. Weapon efficiency will depend on how these times compare: If $t_{crit} > t_{fiss}$, then in principle all of the core material will undergo fission and the efficiency will be 100%. In reality, this is difficult to achieve.

Before proceeding with the detailed analysis, we pause to make a rough estimate of how much time is required to fission the entire core once the chain reaction has been initiated. In Sect. 2.2 we saw that once a neutron is emitted in a fission, it will travel for only about 10 ns before causing another fission. Suppose that we have a core of mass M kilograms of fissile material of atomic weight A grams per mole. The number of nuclei N in the mass will be $N = 10^3 M N_A / A$. If we start with v_o neutrons at the start of "generation 1" and each generation liberates v neutrons, then the algebra of a geometric progression shows that after G generations, the number of fissions F_G that will have occurred, if there is no neutron loss, is $F_G = v_o (v^G - 1) / (v - 1)$. For any sensible value of G, the numerator can be approximated as v^G, and it follows that the number of generations necessary to have fissioned each nucleus is $G = \ln[N(v - 1)/v_o]/\ln(v)$. At τ seconds per generation, it follows that the time to fission the entire mass will be $t_{fiss} \sim \tau G$. For $M = 50$ kilograms of ^{235}U, $N \sim 1.28 \times 10^{26}$. If we set $v_o = 1$ and $v = 2.6$, the number of generations evaluates as $G \sim 63$, which gives $t_{fiss} \sim 0.6$ μs, an incredibly brief time. Even if only half of the neutrons cause fissions ($v = 1.3$), $G \sim 230$, and $t_{fiss} \sim 2.3$ μs. Such are the timescales of nuclear-weapon physics.

Once a chain reaction has been initiated, a bomb core will rapidly (within about a microsecond) heat up, melt, vaporize, and thereafter behave as an expanding gas with the expansion driven by the gas pressure in a thermodynamic $P\Delta V$ manner; we assume that the vast majority of energy liberated in fission reactions can be assumed to go into the form of kinetic energy of the fission products. In what follows, the approach to estimating yield and efficiency will be to use these concepts to establish the range of radius (and hence time) over which the core can expand before the expansion lowers the density of the fissile material to subcriticality. Some fissions will continue to happen after this time, but it is this "criticality shutdown timescale" that fundamentally dictates the efficiency of the weapon.

As above, let τ represent the time for which a neutron will travel before causing a fission; see (2.21). Inverting this, we can say that a single neutron will lead to a subsequent fission at a *rate* of $1/\tau$ per second:

$$rate\ of\ fissions\ per\ neutron = \frac{1}{\tau}. \qquad (2.82)$$

The total number of fissions per second would be this rate times the number of neutrons in the core. The latter will be the product of the number density $N(t) = N_o e^{(\alpha/\tau)t}$ from (2.22) times the volume V of the core. Hence we have

$$fissions/sec = \left(\frac{N_o V}{\tau}\right) e^{(\alpha/\tau)t}. \tag{2.83}$$

In this expression, α is given by solving (2.25), (2.30), and (2.31) for the core at hand, and N_o is the neutron density at the center of the core at $t = 0$; this will be set by the number of neutrons released by some "initiator" device. Recall that $\alpha = 0$ for threshold criticality, whereas $\alpha > 0$ for a core of more than one critical mass, an issue to which we will return shortly.

Equation (2.83) is actually more complicated than it appears, because α and τ are functions of time. To appreciate this, consider a core of some general radius r and density ρ. As the core expands, r will increase while ρ decreases. The decreasing density will cause τ to increase; simultaneously, the discussion following (2.31) indicates that we can expect α to decrease. For sake of simplicity, we assume that α and τ remain constant; not accounting for changes in them will lead to overestimating the fission rate in (2.83). Since an exponential function is involved, the overestimate could be serious; indeed, we will see in Sect. 2.6 that direct use of our resulting yield formula, Eq. (2.95), can easily result in overestimating the efficiency by an order of magnitude. For the present, however, we will stick with the assumption of constant α and τ values since the purpose here is to get a sense of how the expected yield and efficiency depend on the various factors involved. Section 2.6 discusses a simple refinement to (2.83) that eliminates much of the overestimate.

The time required to fission the entire core can be computed by demanding that the integral of (2.83) from time zero to time t_{fiss} to be equal to the total number of nuclei within the core, nV:

$$nV = \left(\frac{N_o V}{\tau}\right) \int_0^{t_{fiss}} e^{(\alpha/\tau)t} dt \Rightarrow t_{fiss} = \left(\frac{\tau}{\alpha}\right) \ln\left[\frac{\alpha n}{N_o}\right], \tag{2.84}$$

where it has been assumed that $e^{(\alpha/\tau)t} \gg 1$ for the timescale of interest, an assumption to be investigated *a posteriori*.

What happens as the exploding core expands? Recall from Sect. 2.2 that the condition for criticality can be expressed as $\rho r \geq K$, where K is a constant characteristic of the material being used. We also saw that for a core of some mass M, $\rho r \propto M/r^2$. As the core expands, the value of ρr must drop, and will eventually fall below the level needed to maintain criticality. This "criticality shutdown" situation will obtain at time t_{crit}, and is technically known as *second criticality*.

For a single critical mass of normal-density material, second criticality will occur as soon as the expansion begins. One way to circumvent this is to provide a tamper to momentarily retard the expansion and so to give the reaction time to build up to a significant degree. Another is to start with a core of more than one critical mass of material of normal density, and this is what is assumed here. The effect of a tamper and the detailed time-evolution of $\alpha(t)$ and τ are dealt with in the following section.

To begin, assume that we have a core of C (> 1) *untamped* threshold critical masses of material of normal density; the initial radius of such a core will be $r_i = C^{1/3} R_o$. (Note that C was used in the preceding section to denote a compression ratio; it now designates a number of critical masses.) We can then solve the diffusion-theory criticality equations, (2.30) and (2.31) for the value of α that just satisfies those equations upon setting the radius to be $C^{1/3}$ times the threshold critical radius listed in Table 2.1.

Now consider the energy released by fissions. If each fission liberates energy E_f, then the rate of energy liberation throughout the entire volume of the core will be, from (2.83),

$$\frac{dE}{dt} = \left(\frac{N_o V E_f}{\tau}\right) e^{(\alpha/\tau)t}. \tag{2.85}$$

Integrating this from time $t = 0$ to some general time t gives the energy liberated to that time:

$$E(t) = \left(\frac{N_o V E_f}{\tau}\right) \int_0^t e^{(\alpha/\tau)t} dt = \left(\frac{N_o V E_f}{\alpha}\right) e^{(\alpha/\tau)t}. \tag{2.86}$$

To determine the pressure within the core, we appeal to a result from thermodynamics. This is that pressure is given by $P(t) = \gamma\, U(t)$, where $U(t)$ is the *energy density* corresponding to $E(t)$: $U(t) = E(t)/V$. The value of the constant γ depends on whether gas pressure ($\gamma = 2/3$) or radiation pressure ($\gamma = 1/3$) is dominant; this issue is discussed below. Thus, the pressure will behave as

$$P(t) = \left(\frac{\gamma N_o E_f}{\alpha}\right) e^{(\alpha/\tau)t} = P_o\, e^{(\alpha/\tau)t}, \tag{2.87}$$

where $P_o = \left(\gamma N_o E_f / \alpha\right)$ is the central pressure at $t = 0$.

The equation of state $P(t) = \gamma\, U(t)$ deserves some comment. In the case of a gas of non-relativistic material particles each of mass m, this expression can be understood on the basis of simple kinetic theory, where one considers the rate at which momentum is transferred to the walls of a container by collisions of the particles with the walls; this is covered in any freshman-level physics or chemistry text. The value of U is taken to be the total kinetic energy of all particles divided by the volume of the container; each particle is assumed to have the same average value of the squared speed, $< v^2 >$. γ emerges from this calculation as 2/3, with the factor of 2 arising from $K = m < v^2 >/2$, and the factor of 3 originating in the presumed isotropy of velocity components over three dimensions. To show that $\gamma = 1/3$ in the case of a gas of photons requires some background in the relativistic energy-momentum relationship of photons, but an ersatz justification can be argued as follows. The non-relativistic result can be re-written as $P = \rho < v^2 >/3$ where ρ is the mass density of the gas. Photons do not have mass, but for the purposes of this quick argument we can use Einstein's famous $E = mc^2$ equation to assign an effective equivalent total mass to the total energy of all photons: $m_{tot} = E_{tot}/c^2$. Hence the density becomes $\rho = m_{tot}/V = E_{tot}/(c^2 V)$, and so the pressure becomes $P = E_{tot} < v^2 >/(3c^2 V)$. Setting $< v^2 > = c^2$, $P = E_{tot}/3 V$, or $P = U/3$ as claimed. In the case of a "gas" of uranium nuclei of standard density of that metal, radiation pressure dominates for per-particle energies greater than about 2 keV (see Exercise 2.14)

How does a gas of photons arise to give a radiation pressure in an exploding bomb core? Fission fragments are bare nuclei and so are highly electrically charged. As they decelerate, they naturally emit energy in the form of photons of wavelengths across the electromagnetic spectrum. Much of the energy released in a nuclear explosion is in the form of gamma-rays and X-rays which ionize the surrounding air.

For simplicity, we model the bomb core as an expanding sphere of radius $r(t)$ with every atom in it moving radially outwards at speed v. Do not confuse this velocity with the average neutron speed v_{neut}, which enters into τ. If the sphere is of density $\rho(t)$ and total mass M, its total kinetic energy will be

$$K_{core} = \frac{1}{2}Mv^2 = \left(\frac{2\pi}{3}\right)\rho v^2 r^3. \qquad (2.88)$$

Now invoke the work-energy theorem in its thermodynamic formulation $W = P(t)dV$, and equate the work done by the gas (or radiation) pressure in changing the core volume by dV over time dt to the change in the core's kinetic energy over that time:

$$\acute{P}(t)\frac{dV}{dt} = \frac{dK_{core}}{dt}. \qquad (2.89)$$

To formulate this explicitly, write, from (2.88), $dK_{core}/dt = (2\pi/3)\rho\, r^3 (2vdv/dt)$, put $dV/dt = 4\pi r^2 (dr/dt)$, and incorporate (2.87) to give

$$\frac{dv}{dt} = \left(\frac{3P_o}{\rho r}\right)e^{(\alpha/\tau)t}. \qquad (2.90)$$

Note that in taking the derivative for dK_{core}/dt, $\rho\, r^3$ can be treated as a constant since it is (but for a factor of $4\pi/3$), the mass of the core. To solve this for the radius of the core as a function of time, we face the problem of what to do about the fact that both ρ and r are functions of time. We deal with this by means of an approximation.

Review the discussion regarding core expansion following (2.83) above. As the core expands, its density when it has any general radius r will be $\rho(r) = C\rho_o(R_o/r)^3$, and criticality will hold until such time as $\rho\, r = \rho_o R_o$, or, on eliminating ρ, $r = C^{1/2}R_o$. We can then define Δr, the range of radius over which criticality holds:

$$\Delta r = r_{\substack{second \\ criticlity}} - r_{initial} = \left(C^{1/2} - C^{1/3} \right) R_o, \tag{2.91}$$

a result we will use shortly.

Now, since $r_i = C^{1/3} R_o$, $(\rho\, r)_{initial} = C^{1/3}(\rho_o R_o)$. For $C = 2$ (for example), this gives $(\rho\, r)_{initial} = 1.26(\rho_o R_o)$. At second criticality we will have $(\rho\, r)_{crit} = (\rho_o R_o)$, so $(\rho\, r)_{crit}$ and $(\rho\, r)_{initial}$ do not differ very greatly. In view of this, we assume that the product ρr in (2.90) can be replaced with a mean value given by the average of the initial and final values of ρr:

$$\langle \rho r \rangle = \frac{1}{2} \left(1 + C^{1/3} \right) \rho_o R_o. \tag{2.92}$$

We can now integrate (2.90) from time $t = 0$ to some general time t to determine the velocity of the expanding core at that time:

$$v(t) = \left(\frac{3P_o}{\langle \rho\, r \rangle} \right) \int_0^t e^{(\alpha/\tau)t} dt = \left(\frac{3P_o \tau}{\langle \rho\, r \rangle \alpha} \right) e^{(\alpha/\tau)t}, \tag{2.93}$$

where it has again been assumed that $e^{(\alpha/\tau)t} \gg 1$.

The stage is now set to compute the amount of time that the core will take to expand through the distance Δr of (2.91). Writing $v = dr/dt$ and integrating (2.93) from $r = r_i$ to $r_i + \Delta r$ for time $t = 0$ to t_{crit} gives

$$t_{crit} \sim \left(\frac{\tau}{\alpha} \right) \ln \left[\frac{\Delta r\, \alpha^2 \langle \rho\, r \rangle}{3 P_o \tau^2} \right] = \left(\frac{\tau}{\alpha} \right) \ln \left[\frac{\Delta r\, \alpha^3 \langle \rho\, r \rangle}{3\,\gamma\,\tau^2 N_o E_f} \right], \tag{2.94}$$

again assuming $e^{(\alpha/\tau)t} \gg 1$ and using $P_o = \gamma\, N_o E_f / \alpha$. Notice that we cannot determine t_{crit} without knowing the initial neutron density N_o. However, since t_{crit} depends logarithmically on N_o, the result is not terribly sensitive to the choice made for that number; presumably the *minimum* sensible value is given by assuming one initial neutron.

The energy yield Y is defined to be the energy released to time t_{crit}. From (2.86) and (2.94), this evaluates as

$$Y = \left(\frac{E_f N_o V}{\alpha} \right) e^{(\alpha/\tau) t_{crit}} = \frac{\Delta r\, \alpha^2 \langle \rho\, r \rangle V}{3\,\gamma\,\tau^2} = \frac{\Delta r\, \alpha^2 \langle \rho\, r \rangle M_{core}}{3\,\gamma\,\tau^2 \rho}. \tag{2.95}$$

Efficiency is defined as the ratio of the yield to the energy which would be liberated if all of the nuclei in the core fissioned:

$$Efficiency = \frac{Y}{E_f\, n\, V} = \frac{\Delta r\, \alpha^2 \langle \rho\, r \rangle}{3\,\gamma\, n\, \tau^2 E_f}. \tag{2.96}$$

Note that the yield and efficiency do not depend on the initial neutron density.

Recall the earlier comments regarding how assuming constant values for α and τ will lead to overestimating the yield; this should be clear by examining (2.96): $1/\tau^2$ will be proportional to the square of the density.

To help determine what value of γ to use, we can compute the total energy liberated to time t_{crit} as in (2.94), and then compute the average energy per particle by dividing by the number of nuclei in the core, nV. The result is

$$\begin{pmatrix} energy\ per\ nucleus \\ at\ time\ t_{crit} \end{pmatrix} = (efficiency)E_f. \tag{2.97}$$

Even if the efficiency is very low, say 0.1%, then for $E_f = 180\,\text{MeV}$ the energy per nucleus would be 180 keV, much higher than the ~2 keV per-particle energy where radiation pressure dominates over gas pressure. It would thus seem reasonable to take $\gamma = 1/3$ in most cases, although $\gamma = 2/3$ would be more appropriate early in the explosion process before much energy has been liberated.

Further, it can be shown by substituting (2.94) into (2.87) and (2.93) that the core velocity and pressure at the time of second criticality are given by

$$v\,(t_{crit}) = \frac{\alpha\,\Delta r}{\tau}, \tag{2.98}$$

and

$$P(t_{crit}) = \frac{\alpha^2\,\Delta r\,\langle \rho r \rangle}{3\,\tau^2}. \tag{2.99}$$

Curiously, this pressure does not depend on the value of γ.

Numbers for uranium and plutonium cores of $C = 1.5$ bare threshold critical masses appear in Table 2.3 (68.8 kg for ^{235}U and 25.04 kg for ^{239}Pu). Secondary neutrons are assumed to have $E = 2$ MeV, and it is assumed that the initial number of neutrons is one.

The timescales and pressures involved in the detonation process are extreme. Criticality shuts down after only 1–2 μs; a pressure of 10^{15} Pa is equivalent to about 10 *billion* atmospheres. Even though $t_{crit}/t_{fiss} \sim 0.9$, the efficiencies are low: small changes in an exponential argument lead to large changes in the results. In the case of ^{235}U, changing the initial number of neutrons to 1000 changes the fission and criticality timescales by only about 10%, down to 1.47 and 1.34 μs, respectively. Also, the comment following (2.84) that $e^{(\alpha/\tau)t}$ can be assumed to be much greater than unity for the timescale of interest can now be appreciated from the fact that $(\alpha/\tau)t_{crit} \sim 50$: $e^{50} \sim 10^{21}$.

Spreadsheet **CriticalityAnalytic.xls** carries out these efficiency and yield calculations for an untamped core. In addition to the parameters already entered for the calculations of the preceding two sections, the user need only additionally specify an initial number of neutrons, a value for γ, and the mass of the core. The "Goal Seek" function is then run a third time, to solve (2.30) and (2.31) for the value of α. The

Table 2.3 Criticality and efficiency parameters for $C = 1.5$, $E_f = 180$ MeV, $\gamma = 1/3$; no tamper. Masses are 68.8 kg for ^{235}U and 25.04 kg for ^{239}Pu

Quantity	Unit	Physical meaning	^{235}U	^{239}Pu
$r_{initial}$	cm	Initial core radius	9.58	7.26
n	10^{22} cm^{-3}	Nuclear number density	4.794	3.930
α	—	Criticality parameter α	0.307	0.376
R_O	cm	Threshold critical radius	8.37	6.345
Δr	cm	Expansion distance to crit shutdown	0.67	0.51
Efficiency	%	Efficiency	1.02	1.29
$P(t_{crit})$	10^{15} Pa	Pressure at crit shutdown	4.72	4.87
Yield	kt	Explosive yield	12.4	5.6
t_{fiss}	μs	Time to fission all nuclei	1.67	1.12
t_{crit}	μs	Time to crit shutdown	1.54	1.04
N_o	Neutron m^{-3}	Initial neutron density	271.8	622.9

Initial number of neutrons = 1
Secondary neutron energy = 2 MeV

spreadsheet then computes and displays quantities such as the expansion distance to second criticality, the fission and criticality timescales, the pressure within and velocity of the core at second criticality, and the efficiency and yield.

When applied to a bare 53 kg core of ^{235}U ($C = 1.155$), **CriticalityAnalytic.xls** indicates that the yield will be about 0.4 kilotons, with a core-expansion distance to second criticality Δr of only 2.1 mm. This yield figure is not directly comparable to the true ~13 kt yield of *Little Boy*, however, as that device was tamped; a more realistic simulation of *Little Boy* that incorporates a tamper is discussed in the next section. The sensitivity of calculations to input parameters is indicated by the fact that if the core mass is raised to 66 kg (the full mass of the *Little Boy* core), the estimated yield rises to ~8.5 kilotons!

How drastically does this analysis tend to overestimate efficiency? In Sect. 2.6 a program is described which carries out a time-dependent simulation of a tamped core. Applying this program to a ^{235}U core of mass 68.8 kg as in Table 2.3 with *no* tamper gives a predicted yield of only about 0.3 kt, about 1/40 of the analytical result of ~12.4 kt. The reason for this drastic discrepancy is explored further at the end of Sect. 2.6. In the meantime, there is a moral here: Beware of the danger of blindly applying an impressive-looking formula.

It is important to emphasize that the above calculations cannot be applied to a tamped core; that is, one cannot simply solve (2.47) and (2.48) for a core of some specified mass and tamper of some outer radius and use the value of α so obtained in the time and efficiency expressions established above. The reason for this has to do with the distance Δr through which the core expands before second criticality, Eq. (2.91). This expression derived from the fact that the criticality equation for the untamped case involves the density and radius of the core in the combination $\rho\, r$; in the tamped case the criticality condition admits no such combination of parameters, so the subsequent calculations of criticality timescale and efficiency do not transform unaltered to using a tamped core. Efficiency in the case of a tamped core can only be established numerically.

The conditions that exist within a nuclear explosion are so fantastic by everyday standards that simply manipulating formulae or running specific calculations cannot really give an overall sense of the orders of magnitude involved. To this end, I develop here an approximate graphical representation of the evolution of neutron density, energy density, pressure, and temperature within an exploding bomb core as a function of time. This approach is based on a hand-drawn graph prepared in early 1943 by Robert Serber for his *Los Alamos Primer*. Serber's goal was not a precise analysis—many nuclear parameters were only approximately known at the time—but rather to try to get across to his colleagues a sense the extreme conditions that they would be dealing with. Serber's plot still stands as a remarkable example of effective graphical display of information. The development given here is adapted from previous publications (Reed 2016, 2020c).

From Eqs. (2.22), (2.86), and (2.87), the neutron density N, energy density U, and pressure P within an exploding bomb core can be written as

$$N(t) \;=\; N_o e^{(\alpha/\tau)t}, \tag{2.100}$$

$$U \;=\; \left(\frac{E_f}{\alpha}\right) N_o e^{(\alpha/\tau)t}, \tag{2.101}$$

and

$$P \;=\; \gamma\, U. \tag{2.102}$$

In early 1943, Serber had no idea what value α would take for an eventual bomb design, but knew it would likely be on the order of a few tenths by virtue of Eq. (2.34). Similarly, the neutron travel time τ depends on the fission cross-section; while this was still being pinned down experimentally, it was already estimated to be on the order of ~ 10 ns. For simplicity, Serber took $\alpha = 1$, and used τ as his unit of time. The neutron density is of course also a function of position within the core, but Serber's concern was the order of magnitude of the time- dependence of the various quantities involved. Even though radiation pressure dominates over gas pressure in the later stages of the explosion, I follow Serber and adopt $\gamma = 2/3$; for the purpose intended here, a factor of two is of no real consequence.

Serber did not include any indication of temperature in his graph; I do so by assuming that the vaporized fissile material acts as an ideal gas, in which case temperature can be determined through the pressure via

$$P_{gas} \;=\; n\,k\,T. \tag{2.103}$$

Serber had in mind using ^{235}U, for which $n \sim 4.794 \times 10^{28}$ m^{-3}.

The graph is shown in Fig. 2.15. Following Serber, I adopted $N_o = 10^5$ cm^{-3}; he largely used cgs units. The choice of N_o is entirely arbitrary, affecting only the location of the zero-point of the horizontal axis in the graph. Pressure is plotted in atmospheres. Plotting separate graphs for pressure and energy density is essentially redundant, but doing so allows convenient units to be invoked for each. Efficiency is computed on the basis of comparing the energy density as a function of time to that which would obtain if all of the uranium nuclei in one cubic meter of material were fissioned, which for $E_f = 170$ MeV (what Serber adopted) corresponds to 1.31×10^{18} J m^{-3}.

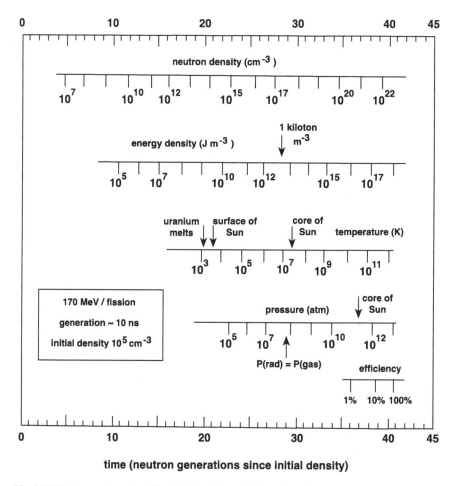

Fig. 2.15 Elaborated version of Serber's *Los Alamos Primer* plot of conditions inside an exploding nuclear weapon. This is not intended to depict the results of precise calculations, but rather to indicate orders of magnitude

The orders of magnitude involved in Fig. 2.15 drive home the remarkable conditions that prevail during nuclear explosions, and how the vast majority of the energy is liberated during only the last few fission generations. With ~10 ns per generation, the entire timescale the figure is only about one-half of a microsecond. Particularly striking is the rate of growth of temperature: a mere 10 generations elapse between when this reaches that of the surface of the Sun and that at its core.

The computations involved in preparing Fig. 2.15 are so trivial that they make it easy to overlook just how remarkable this plot is. The results are largely independent of any particular bomb design, core mass, critical mass, fissile material, or number of neutrons per fission. The graph gets across the orders of magnitude involved in as simple and general a way as one could desire without undue sacrifice of the relevant physics. Further, a psychological aspect of the way it is laid out deserves comment. Usually we plot exponential functions using logarithmic scales so that the relationships appear as straight lines. Everything in Fig. 2.15 could be presented this way, but this would have the effect of downplaying the magnitudes of some of the quantities involved. For example, the greatest neutron density involved is 10 orders of magnitude greater than that of the greatest pressure; in a logarithmic plot, the line for the former will always lie well above that of the latter and so draw the lion's share of one's attention. A density of a few quadrillion neutrons per cubic centimeter sounds fantastic, but in actuality corresponds to fission of less than one part in a million of the ^{235}U contained in any one cubic centimeter of the core. In contrast, by the time that the neutron density has reached this value, the temperature is greater than that of the surface of the Sun, and the pressure has risen to ~100,000 atmospheres. These latter numbers are sure to impress anybody.

The values indicated in Fig. 2.15 agree closely with what Serber gives, with one exception. He states that the gas and radiation pressures will be equal at ~36 generations, but it is not clear how he arrives at this value. For an environment at absolute temperature T, thermodynamics gives an expression for the radiation pressure:

$$P_{rad} = \left(\frac{8\pi^5 k^4}{45c^3h^3} \right) T^4 = \left(2.524 \times 10^{-16} \ Pa \ K^{-4} \right) T^4, \qquad (2.104)$$

where the various symbols have their usual meanings. Adopting n as above for ^{235}U, the gas and radiation pressures will be equal at $T \sim 1.38 \times 10^7$ K and $P \sim 9.14 \times 10^{12}$ Pa, which (2.102) indicates will occur at $t \sim 29$ generations ($\gamma = 2/3$). This may simply have been a computational error on Serber's part in the haste of preparing his lectures.

To close this section, we compare the efficiency formula derived here to what was probably the first recorded formulation of the energy expected to be liberated by a nuclear weapon. This appeared in a document which has come to be known as the *Frisch-Peierls Memorandum*. This remarkable 7-page manuscript was prepared by Otto Frisch and Rudolf Peierls in March, 1940, to alert British government and military officials to the possibly of creating extremely powerful bombs based on utilizing fission; the title of their memo was "On the construction of a "super-bomb", based on a nuclear chain reaction in uranium." Their work was remarkably prescient: They discussed how a chain reaction could not happen in ordinary uranium; raised the possibility of bringing together two subcritical pieces of pure ^{235}U to create a supercritical mass; discussed how neutrons in cosmic radiation could be used to trigger the device; described how ^{235}U could be isolated by diffusion; and remarked that such a device would create significant radioactive fallout. Copies of the memorandum can be found in many online sites; a printed copy appears in Serber (1992). Readers are warned, however, that many reprintings contain various typographical errors. A detailed analysis of the physics involved in the memorandum is presented by Bernstein (2011), who also describes the errors.

The only mathematical expression appearing in the Frisch-Peierls memorandum is one for the expected yield of an untamped weapon. In terms of the notation of this book, this appears as

$$Y = 0.2 M_{core} \left(R_{core} / \tau \right)^2 \left(\sqrt{R_{core} / R_o} - 1 \right). \tag{2.105}$$

This looks almost completely unlike the present yield formula, (2.95). However, the latter can be transformed into (2.105) in a few steps via some sensible approximations. First, write the core volume or mass in (2.95) in terms of the core radius; also, set $\gamma = 1/3$. These manipulations give

$$Y = \frac{4 \pi R_{core}^3 \alpha^2 \Delta r \langle \rho r \rangle}{3 \tau^2}. \tag{2.106}$$

Now consider the product $\Delta r \langle \rho r \rangle$. From (2.91) and (2.92),

$$\Delta r \langle \rho r \rangle = \frac{1}{2} \left(C^{1/2} - C^{1/3} \right) \left(1 + C^{1/3} \right) \rho_o R_o^2. \tag{2.107}$$

In the second bracket in this expression, make the approximation that $C^{1/3} \sim 1$ to give $(1 + C^{1/3}) \sim 2$. This is reasonable as that bracket contains the *sum* of two similar quantities. We do not make this approximation within the first bracket, however, as it contains the *difference* of two similar quantities. In this case, extract a factor of $C^{1/3}$ from within the bracket, and write it as $C^{1/3} = R_{core}/R_o$. The factor of $C^{1/6}$ remaining within the first bracket can then be written as $\sqrt{R_{core} / R_o}$. Thus, (2.107) becomes $\Delta r \langle \rho r \rangle \sim \left(\sqrt{R_{core} / R_o} - 1 \right) \rho_o R_o R_{core}$. On substituting this into (2.106), we can write $4\pi R_{core}^3 \rho_o / 3 = M_{core}$, and the yield becomes

$$Y \sim \alpha^2 M_{core} \left(R_o R_{core} / \tau^2 \right) \left(\sqrt{R_{core} / R_o} - 1 \right). \tag{2.108}$$

Finally, it is not unreasonable to make the approximation $R_{core}\, R_o \sim R_{core}^2$, and so arrive at

$$Y \sim \alpha^2 M_{core} \left(R_{core}/\tau\right)^2 \left(\sqrt{R_{core}/R_o} - 1\right), \qquad (2.109)$$

precisely the form of the Frisch-Peierls formula. They evidently took $\alpha^2 = 0.2$. On considering that we just found $\alpha_{initial} = 0.307$ for 1.5 critical masses of ^{235}U, their estimate was reasonable if a little on the high side. Frisch and Peierls must have worked out the relevant diffusion and criticality theory "in the background" before composing their memorandum. Peierls was a master theoretical physicist, very familiar with diffusion problems; in Sect. 2.7 we will examine a formulation of criticality that he had published in the fall of 1939, several months before he teamed up with Frisch to produce their now-famous memorandum.

2.6 Estimating Yield—Numerical

In this section, a numerical approach to estimating weapon efficiency and yield is developed. The essential physics necessary for this development was established in the preceding sections; what is new here is how that physics is used. The analysis presented in this section is adopted from Reed (2010).

The approach taken here is one of standard numerical integration: The parameters of a bomb core and tamper are specified, along with a timestep Δt. At each timestep, the energy released from the core is computed, from which the acceleration of the system at that moment can be determined. The expansion of the system is tracked until second criticality occurs. The tamper is assumed to have the same expansion speed as the core, so their radial expansions will be the same.

The integration process involves eight steps:

(i)   Fundamental parameters are specified: The mass of the core, its atomic weight, initial density, and nuclear characteristics σ_f, σ_{el}, and ν. Similarly, the mass, atomic weight, density, and elastic-scattering cross-section of the tamper are specified. The energy release per fission E_f and gas/radiation pressure constant γ are set. A timestep Δt also needs to be chosen; this is discussed below. The initial number of neutrons also has to be specified, since this value enters into the fission rate and energy release at each timestep in steps (iv) and (v) below.

(ii) Elapsed time, the expansion speed of the system, and the total energy released are initialized to zero. Core and tamper radii are initialized according as their masses and densities.

(iii) The exponential neutron-density growth parameter α is determined by numerical solution of (2.47) and (2.48).

(iv) The rate of fissions at a given time is computed from (2.83):

$$fissions/sec = \left(\frac{N_o V_{core}}{\tau} \right) e^{(\alpha/\tau)\, t}. \tag{2.110}$$

Only the core volume is used in computing the fission rate, as the tamper is assumed to be non-fissile.

(v)   The amount of energy released during time Δt is computed from (2.85):

$$\Delta E = \left(\frac{N_o V_{core} \, E_f}{\tau} \right) e^{(\alpha/\tau)\, t} (\Delta t). \tag{2.111}$$

(vi) The total energy released to time t is updated, $E(t) = E(t) + \Delta E$, and, from the discussion following (2.86), the pressure at time t is given by

$$P_{core}(t) = \frac{\gamma \, E(t)}{V_{core}(t)}. \tag{2.112}$$

I use the core volume here on the rationale that the fission products which cause the gas/radiation pressure will likely largely remain within the core.

(vii) A key step is computing the change in the expansion speed of the core and tamper over the elapsed time Δt due to the energy released during that time. In the discussion leading up to (2.89), this was approached by invoking the work-energy theorem:

$$P(t) \frac{d V_{core}}{dt} = \frac{d K}{dt}. \tag{2.113}$$

To improve the veracity of the simulation, it is desirable to account, at least in some approximate way, for the retarding effect of the tamper on the expansion of the core. To do this, I treat the dK/dt term as involving the sum of the core and tamper masses. The dV/dt term is taken to apply to the core only. With r as the radius and v the speed of the core, we have

$$\frac{\gamma \, E(t)}{V_{core}(t)} \left(\frac{d V_{core}}{dt} \right) = \frac{d K_{total}}{dt}$$

$$\Rightarrow \frac{\gamma \, E(t)}{V_{core}(t)} \left(4\pi \, r^2 \frac{dr}{dt} \right) = \frac{1}{2} M_{total} \left(2v \frac{dv}{dt} \right).$$

From this, we can compute the change in expansion speed of the core over time Δt as

$$\Delta v = \left[\frac{4\pi \, r^2 \gamma \, E(t)}{V_{core} \, M_{total}} \right] (\Delta t). \tag{2.114}$$

With this, the expansion speed of the system core and the core and tamper radii are updated according as $v(t) = v(t) + \Delta v$ and $r(t) = r(t) + v(t)\Delta t$.

(viii) Increment time according as $t = t + \Delta t$, and return to step (iii) to begin the next timestep; continue until second criticality is reached when $\alpha = 0$. At the beginning of each timestep, the core and tamper densities must be updated, as well as their nuclear number densities, mean free paths, and the neutron travel time between fissions.

The assumption that the core and tamper experience the same expansion speed is a quite arbitrary one for sake of simplicity of the programming. Other assumptions could be made (such as, perhaps, having the tamper retain a constant density), none of which are likely to be particularly realistic. Nuclear engineers speak of the "snowplow" effect, where high-density tamper material piles up outside the expanding core/tamper interface. But the point here is an order-of-magnitude pedagogical model. Also, this same assumption is made in a simplified yield model developed in Sect. 2.8.4; this allows for some cross-checking of results.

What of the timestep Δt? In setting this, it is helpful to appreciate that it is not necessary to start a simulation at $t = 0$. From (2.111), little energy will be released while $(\alpha/\tau)t$ is small. An example using ^{235}U will help make this clear. With $\tau \sim 8.64 \times 10^{-9}$ s (Table 2.2). and, say, $\alpha \sim 0.5$, then $(\alpha/\tau) \sim 5.8 \times 10^7$ s^{-1}. Starting a simulation at $t = 10^{-8}$ s should thus sacrifice no accuracy. However, the choice of a timestep Δt is a sensitive issue, since the rate of energy release grows exponentially at later times. For a function of the form $y = \exp[(\alpha/\tau)t]$, the fractional change in y over a time Δt will be $dy/y = (\alpha/\tau) \Delta t$; to have dy/y be small suggests adopting a value of Δt no larger than the inverse of (α/τ), which is about 1.7×10^{-8} s. Consequently, all of the results described in what follows utilized a starting time of 10^{-8} s, and a timestep of $\Delta t = 5 \times 10^{-10}$ s; a run to a final time of 1.1 microseconds will involve nearly 2,200 timesteps. With this value of Δt, $dy/y \sim 0.029$.

This author has developed a FORTRAN program for carrying out this simulation; the code and an accompanying user manual are available upon request.

2.6.1 A Simulation of the Hiroshima Little Boy Bomb

Figures 2.16 and 2.17 show the results of a simulation of an idealized *Little Boy* configuration: a 53 kg ^{235}U core plus a 550 kg WC tamper. The initial core radius is 8.78 cm, and the initial outer radius of the tamper is 20.86 cm. The initial number of neutrons was set to be one. Clearly, the exponential parameter α remains largely unchanged during the time over which the bulk of the energy is emitted. Second criticality occurs when the core has expanded to a radius of 11.08 cm at a time of 1.18 μs, an expansion distance of $\Delta r = 2.3$ cm. As remarked earlier, a tamper significantly affects the expansion distance over which criticality holds. The total yield is estimated at 11.95 kt; this is discussed further below.

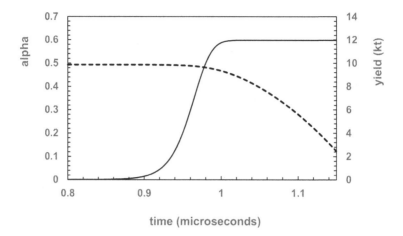

Fig. 2.16 Exponential growth parameter α (dashed line) and yield (kilotons; solid line) versus time for a simulation of the *Little Boy* bomb: 53 kg core plus 550 kg tungsten-carbide tamper

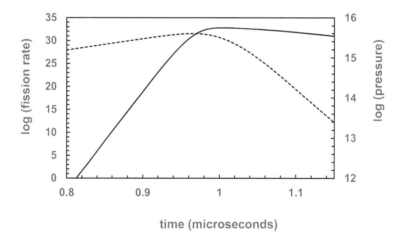

Fig. 2.17 Logarithm (base 10) of fission rate (dashed curve, left scale) and logarithm of pressure in Pa (solid curve, right scale) versus time for a simulation of the *Little Boy* bomb

Figure 2.17 shows the runs of fission rate and pressure as functions of time. The pressure peaks at about 5.6×10^{15} Pa, or about 56 *billion* atmospheres. The fission rate peaks at about 3×10^{31} per second. At second criticality, the core expansion velocity is about 200 km s^{-1}. These graphs dramatically illustrate what Robert Serber wrote in *The Los Alamos Primer*: "Since only the last few generations will release enough energy to produce much expansion, it is just possible for the reaction to occur to an interesting extent before it is stopped by the spreading of the active material."

The yield as estimated by this simulation is in reasonable agreement with published estimates. A 1952 Los Alamos report on the Hiroshima bombing, http://www.fas.org/sgp/othergov/doe/lanl/la-1398.pdf, estimated 18.5 $\pm$ 5 kt. A later analysis published by Penney et al. (1970) estimated ~12 kt, and a 1985 Los Alamos reassessment indicated ~15 kt (Malik 1985). The simulation assumes a spherical core, not a cylindrical one, which is likely to be less efficient; however, a 53 kg core at a density of 18.71 gr cm^{-3} would have a diameter of ~6.9 inches, very close to the 7- inch true length of the *Little Boy* cylinder, which was almost as wide as it was long: see Fig. P.3 in the Preamble. The simulation takes no account of any fissioning of ^{238}U by high-energy neutrons, and so in this sense is likely to return an underestimate. A moral here is that nuclear explosions are notoriously difficult to model accurately. At a fission yield of 17.6 kt per kg of pure ^{235}U (180 MeV/fission), the simulation estimate represents an efficiency of about 1.3% for the 53-kg core.

Figure 2.18 shows how the simulated yield of a 53-kg core varies as a function of tamper mass. A linear fit indicates that each additional kilogram of tamper increases the yield by ~0.023 kt. Of course, we would expect this curve to eventually level off to the theoretical maximum yield as the tamper mass becomes very great.

It was remarked in Sect. 2.4 that a simulation of an *untamped* ^{235}U core of mass 68.8 kg ($C = 1.5$ bare critical masses) results in a yield of only 0.3 kt, about 1/40 that predicted by Eq. (2.95). Why are these predictions so wildly discrepant? The culprit proves to be that in deriving (2.95), the exponential factor α was assumed to be constant. Look back to Fig. 2.16, which shows that once α begins to decline, very little additional yield occurs. In assuming that α remains constant until the core

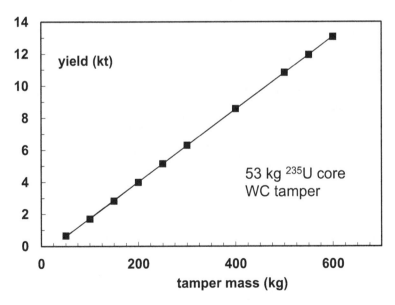

Fig. 2.18 Yield of a 53-kg ^{235}U core versus mass of surrounding tungsten-carbide tamper. The line is interpolated

reaches second criticality, (2.95) consequently seriously overestimates the yield. Some numbers for the 68.8 kg simulation are instructive. The initial core radius in this case is 9.575 cm, and the initial value of α is 0.3096. The second-criticality radius is 10.255 cm ($\Delta r = 0.68$ cm), but by the time that the radius has expanded to only 9.607 cm (an increase of only 0.336%), fully 90% of the final yield has already been realized. By this time, α has dropped by only about 4.5% from its initial value, but the reaction has already begun shutting down. (It is true that Fig. 2.16 is a tamped-core simulation, but the behavior of α is very similar for an untamped core.)

Can (2.95) be modified to account for this problem? Here is a possible approach: When integrating (2.93) to determine the time of second criticality, replace the upper limit of integration $r_i + \Delta r$ with $r_i(1 + f)$, where f is the fractional increase in the core radius corresponding to that time at which you think the reaction begins shutting down; for example, for the above numbers, $f = 0.0034$ corresponds to 90% energy release. Carrying out the integral shows that the yield emerges as (2.95) except that the factor of Δr in the numerator is replaced with fr_i. For the present case of $r_i = 9.575$ cm and $f = 0.00336$, this modification predicts a yield of 0.597 kt, just twice the simulation result. There is obviously no preferred value of f to use, but this artifice removes much of the discrepancy in a straightforward way.

The same program was used to simulate a 6.3 kg core of ^{239}Pu surrounded by a 230 kg tamper of depleted uranium, with both compressed to densities 2.5 times as great as their normal densities to account for the implosion nature of this weapon. (230 kg is the total mass of the *Trinity* tamper shells as described in the Preamble.) In this case, the estimated yield is 17.2 kt, which at first glance seems in only fair agreement with an estimated true yield of 21 kt (Malik 1985). However, since some 30% of *Trinity's* yield was contributed by high-energy neutrons inducing fissions in the DU tamper shell (that is, ~6 kt; Semkow et al. 2006), ~17 kt is reasonable for the plutonium-only fraction of the yield. The results of these simulations should not be over-interpreted, but it is encouraging to see that the model gives results of the correct order of magnitude.

To close this section, a dose of perspective: Do not be *too* upset that Eq. (2.95) is not very accurate. It pertains to an untamped core, and any serious bomb-maker will incorporate a tamper. Ultimately, numerical simulations are what tell the tale of efficiency and yield. Also, treat discrepancies as valuable lessons: Analytic results have a compelling attractiveness and are powerful for getting a sense of how something depends on the parameters involved, but always be prepared to question underlying assumptions.

2.7 History Lesson: Criticality Considered in 1939

In Sect. 2.2, we saw that the criticality condition for threshold criticality ($\alpha = 0$) for an untamped core can be expressed as (Eqs. 2.30 and 2.31)

$$x \cot(x) + \varepsilon x - 1 = 0, \tag{2.115}$$

with

$$\varepsilon = \frac{1}{2}\sqrt{\frac{3\lambda_f}{\lambda_t(\nu - 1)}} = \frac{1}{2}\sqrt{\frac{3\sigma_t}{\sigma_f(\nu - 1)}}. \tag{2.116}$$

Once the nuclear parameters σ_f, σ_{el}, and ν are set, ε is determined, and the solution of (2.115) for x can be plotted as a function of ε as in Fig. 2.5. For a given value of $x(\varepsilon)$, the critical radius R follows from (2.26):

$$R = d\, x = \sqrt{\frac{\lambda_f \lambda_t}{3(\nu - 1)}}\, x = \frac{1}{n}\sqrt{\frac{1}{3\sigma_f\,\sigma_t(\nu - 1)}}\, x. \tag{2.117}$$

As formulated, (2.115) and (2.116) are convenient in that both x and ε are dimensionless, but they are awkward in that ε is not bounded: If ν is very large, ε will approach zero, but if $\nu \to 1$, ε diverges to infinity. It would be handy to have some combination of σ_f, σ_{el}, and ν that is finitely bounded. Such a combination was developed by Peierls (1939) in a paper which was the first publication in English to explore what he termed "criticality conditions in neutron multiplication." He defined a dimensionless quantity ξ given by

$$\xi^2 = \frac{\sigma_f(\nu - 1)}{\sigma_{el} + \nu\sigma_f}. \tag{2.118}$$

For $1 \le \nu \le \infty$, $0 \le \xi \le 1$. Note that it is the elastic-scattering cross-section σ_{el} that appears in the denominator of this definition, rather than transport cross-section $\sigma_t = \sigma_{el} + \sigma_f$.

If $(\nu - 1)$ is eliminated between (2.116) and (2.118), ε and ξ prove to be related as

$$\varepsilon = \sqrt{\frac{3}{4}\left(\frac{1}{\xi^2} - 1\right)}. \tag{2.119}$$

Similarly, if $(\nu - 1)$ is extracted from the definition of d in (2.117) and substituted into (2.118), then one finds

$$d = \sqrt{\frac{1}{3}\left(\frac{1}{\xi^2} - 1\right)}\,\lambda_t. \tag{2.120}$$

A general formulation of critical radii can now be made as follows: For a range of values of ξ between zero and one, (2.115) and (2.119) can be solved for x. For each solution, (2.117) and (2.120) show that the ratio of R to λ_t can be expressed purely as a function of ξ:

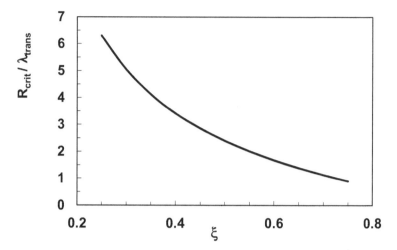

Fig. 2.19 Ratio of untamped threshold critical radius to transport mean free path as a function of Peierls' ξ parameter of (2.118)

$$\frac{R}{\lambda_t} = x(\xi)\,d(\xi) = x(\xi)\sqrt{\frac{1}{3}\left(\frac{1}{\xi^2} - 1\right)}. \qquad (2.121)$$

In other words, a graph of $x(\xi)\,d(\xi) \equiv R/\lambda_t$ versus ξ can be used to immediately indicate the ratio of the untamped threshold critical radius to the transport mean free path for any combination of σ_f, σ_{el}, and ν values. As with Fig. 2.5, the advantage of this approach is that the graph need only be constructed once.

Figure 2.19 shows R/λ_t as a function of ξ. For ^{235}U and ^{239}Pu, $\xi \sim 0.5084$ and 0.6221, and $R/\lambda_t \sim 2.33$ and 1.54, respectively. It is intuitively sensible that for small values of ξ (that is, for $\nu \to 1$), the critical radius will be large, and vice versa.

An important aspect of Peierls' analysis is that it provides an independent check on the diffusion method of analyzing critical mass that has been used throughout this chapter. Peierls showed that his analysis led to approximate analytic solutions for the critical radius R in two limiting cases: $\xi \to 0$ and $\xi \to 1$. These are given by

$$\frac{1}{\beta R} \sim \begin{cases} 0.552\xi + 0.216\xi^2 & (\xi \to 0) \\ 0.78 - 1.02(1-\xi) & (\xi \to 1), \end{cases} \qquad (2.122)$$

where

$$\beta = n\left(\sigma_{el} + \nu\sigma_f\right). \qquad (2.123)$$

β is identical to the denominator of (2.118) but for a factor of the nuclear number density n.

βR can be expressed in terms of x and ξ through the following manipulations. First, from (2.119) and (2.120), we can write $d = 2\lambda_t \varepsilon/3$. With this result, we can write $x = R/d$ as $x = 3R/(2\lambda_t \varepsilon)$. By eliminating σ_f times $(\nu - 1)$ between (2.116) and (2.118), we get $\lambda_t = 3/(4\beta\xi^2\varepsilon^2)$. Substituting this result into the expression for x then shows that

$$\frac{1}{\beta R} = \frac{2\xi^2\varepsilon}{x}. \tag{2.124}$$

We can compare the results of Peierls' approach to those of diffusion analyses in much the same way as Fig. 2.19 was constructed: For a range of values of ξ between zero and unity, solve (2.115) and (2.119) for x, which can be translated to $1/(\beta R)$ through (2.124) and then compared to the predictions of (2.122). Figure 2.20 shows the results of such an analysis for $0.1 \le \xi \le 0.9$. It is reassuring to see that the results of the diffusion analysis do not differ markedly from those of Peierls'. This is particularly true for small values of ξ, where the core will be large and we expect diffusion theory to be accurate; curiously, the diffusion approach *overestimates* the critical radius for $\xi \to 1$. For ^{235}U, (2.122) predicts critical radii of 7.93 cm ($\xi \to 0$) and 9.57 cm ($\xi \to 1$). These radii correspond to masses of 39–69 kg, which bracket the diffusion result of 46 kg. For ^{239}Pu, the Peierls-method masses evaluate as 13.4 and 17.0 kg, which again bracket the diffusion result of 16.7 kg.

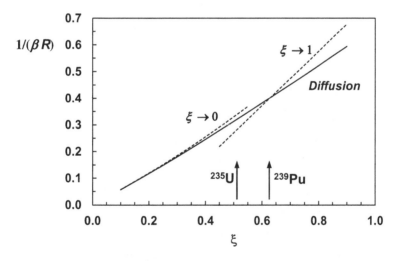

Fig. 2.20 $1/(\beta R)$ computed with Peierls' limiting expressions of (2.122) (dashed lines) and diffusion analysis (solid line) versus Peierls' ξ parameter of (2.118). Adopted from Reed (2008)

2.8 Criticality and Yield: Approximate Methods

The approximate methods described in the following subsections vary from extremely straightforward algebraic ones to more sophisticated treatments involving integration and numerical root-finding. These are not intended to replace the more rigourous analyses of previous sections, but rather to provide some "back of the envelope" techniques for making quick estimates and for showing how quantities depend on the fundamental parameters involved.

2.8.1 Bare Critical Mass: Simplified Boundary Condition

In Sect. 2.2, the neutron-flux boundary condition (2.29) resulted in a transcendental equation, (2.30), for the criticality condition. A simpler result can be had by relaxing the boundary condition to assume that the density of neutrons drops to zero at the edge of the core: $N(R_C) = 0$. This results in an overestimate of the critical radius because the demand of no neutron density is too pessimistic in that it corresponds to not permitting any loss of neutrons from the core, but has the advantage of not requiring any numerical root-finding. In this case, the solution for the neutron density, (2.28), will satisfy the boundary condition when $x = \pi$, which means, via (2.26), that $R_C = xd = \pi\, d$. Assuming threshold criticality with $\alpha = 0$, this gives, with (2.25)

$$R_{crit} = \pi \sqrt{\frac{\lambda_f\, \lambda_t}{3\,(\nu - 1)}} = \frac{\pi}{n} \sqrt{\frac{1}{3\,(\nu - 1)\sigma_f\, \sigma_t}}. \qquad (2.125)$$

For ^{235}U and ^{239}Pu, this gives $R_{crit} = 11.05$ cm and 9.38 cm, corresponding to masses of about 106 kg and 54 kg, respectively. These are about 2.3 and 3.2 times the diffusion-theory values. The value of this approach, which was used by Robert Serber in the *Los Alamos Primer,* is that it makes very clear how the critical mass depends on the nuclear parameters σ_f, σ_t, ν, and n.

2.8.2 Bare Critical Mass: An Even Simpler Approach

This section presents an even simpler approach to estimating the critical mass for a bare core. This approach is straightforward enough to be presented to, say, an upper-level high-school class, provided that the students have been introduced to concepts such as cross-sections, secondary neutrons, and the empirical escape expression $N_{esc} = N_o e^{-\sigma n x}$ of Sect. 2.1. This development is adapted from Reed (2018a).

Imagine again a spherical sample of fissile material. Let the total number of nuclei in the sample be N, and again denote the average number of secondary neutrons emitted per fission as ν. If all nuclei are to fission, then νN secondary neutrons will ultimately be created. All νN secondary neutrons will either escape or be consumed in causing fissions: $\nu N = N_{escape} + N_{fission}$. If we take the condition to have a critical mass to be that each nucleus does in fact fission, then N of the secondary neutrons must be consumed in causing fissions. Under this assumption we can write $\nu N = N_{escape} + N$, or $N_{escape} = N(\nu - 1)$. This is the number of neutrons that we can *permit* to escape. However, we know from the development in Sect. 2.1 that Eq. (2.7), $N_{esc} = N_o e^{-\sigma n x}$, dictates how many *must* escape, when modified to represent a spherical core in place of a linear structure of thickness x. The condition for a critical mass then becomes demanding that the number that must escape be less than or at most equal to the number that can be permitted to escape:

$$N_{\substack{must \\ escape}} \leq (\nu - 1)N. \tag{2.126}$$

The issue of modifying N_{esc} for a spherical core is examined in detail in Sect. 6.6 and Exercise F.1, where it is shown that the average distance from any point within a sphere or radius R to its surface is $3R/4$. If we assume that fission is the *only* reaction that can happen when a nucleus is struck by a fleeing neutron, then we have

$$N_{escape} = \nu N \exp\left(-\frac{3}{4}\sigma_f n R\right). \tag{2.127}$$

Combining this with (2.126) and solving for the radius R gives

$$R_{crit} \sim \frac{4}{3\,\sigma_{fiss}\,n} \ln\left(\frac{\nu}{\nu - 1}\right), \tag{2.128}$$

which is written as an approximation to emphasize that scattering has been neglected.

This expression gives critical radii of 10.74 and 7.14 cm for ^{235}U and ^{239}Pu, respectively, corresponding to masses of about 97 and 24 kg. These results are closer to the formal diffusion theory results than those derived in the preceding section, presumably because they allow for some neutron escape. Sometimes, a simpler approximation turns out to be a better one!

2.8.3 Estimating the Yield of the Trinity Test by Examining the Rate of Growth of the Fireball

In this section, the order of magnitude of the explosive yield of the *Trinity* test is estimated through a high-school-level conservation of energy analysis of the growth of its fireball. This analysis is adapted from Reed (2020b), and is somewhat akin to the treatment of bomb efficiency in Sect. 2.4. Figure 2.21 shows the *Trinity* fireball at 25 ms after the explosion. A series of time-lapse images of the explosion can be found at https://www.trinityremembered.com/photos/test/.

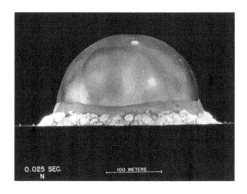

The characteristics of the shock wave formed by the sudden release of a great
amount of energy in a small volume were studied theoretically by British physicist
Geoffrey Taylor in a secret report prepared in 1941. Taylor's analysis was dauntingly
complex, involving advanced differential equations to treat the thermodynamics of
highly ionized and compressed air. The net result, however, was a prediction that in
its initial stages, the fireball radius r should grow as the two-fifths power of elapsed
time: $r \propto t^{2/5}$. Taylor's analysis was published in 1950 along with a companion
paper which analyzed the growth of the *Trinity* fireball as deduced from declassified films of the explosion (Taylor 1950a, b). His data comprised measurements of
the fireball radius over the time span 0.10–62 ms following the explosion, and he
found, somewhat to his surprise in view of approximations invoked in his original
analysis, that the two-fifths-power law held remarkably closely over this span. That
the predicted power law held so well is even more surprising in that the fireball hit
the ground within a millisecond of the explosion; this must have absorbed some of
the available energy.

Empirically, Taylor found that for r in meters and t in seconds, the fireball radius
could be expressed as

$$\log(r) \;=\; 2.766 \;+\; \frac{2}{5}\log(t), \tag{2.129}$$

or

$$r \;=\; 583.5 \, t^{2/5} \tag{2.130}$$

As is described in what follows, the two-fifths power dependence can be "derived"
via a straightforward conservation of energy argument. The empirical factor of 583.5
can then be used to estimate the energy of the explosion.

When a nuclear weapon is detonated, an enormous amount of energy is released.
According to Glasstone and Dolan (1977), results gleaned from years of nuclear tests
indicate that ~50% of the energy yield goes into a shock wave of compressed air that
spreads outward from the explosion, with the remainder distributed between thermal

radiation (~35%), prompt ionizing radiation (~5%), and longer-term residual fallout radiation (10%); these authors also give a detailed description of fireball formation and evolution. Since the nuclear explosion itself takes only about one microsecond, all of the energy released in the *Trinity* explosion was emitted well before the 0.1 ms initial time of Taylor's analysis; we can assume that no further energy is generated during the span of his data. For the present purpose, I will assume that an amount of energy K goes into the kinetic energy of a bubble of air which expands outward, accumulating more air as it does so. After securing an estimate of K, the overall yield of the bomb can be estimated by multiplying by a factor of two to account for the shock-wave share of the total energy as indicated above.

If the mass of the air bubble at any time is m, then $K = mv^2/2$, where the expansion speed v is (dr/dt). If the density of air is ρ, then when the bubble has radius r we must have $m = 4\pi\, r^3\rho/3$. Hence, at any moment,

$$K = \frac{1}{2}\left(\frac{4}{3}\pi r^3\rho\right)\left(\frac{dr}{dt}\right)^2. \tag{2.131}$$

Take the square root of this expression and separate variables to give

$$\sqrt{\frac{3K}{2\pi\rho}}\, dt = r^{3/2}dr. \tag{2.132}$$

Integrating from $r = 0$ at $t = 0$ to some general later time t gives

$$r = \left(\frac{5}{2}\sqrt{\frac{3K}{2\pi\rho}}\right)^{2/5} t^{2/5}. \tag{2.133}$$

While the 2/5 power might be argued as to be expected on the basis that energy must be conserved no matter how complex a phenomenon, it seems surprising that we can recover Taylor's radius-time behavior with such a simple argument; after all, the fireball will be a complex soup of bomb debris, fission products, photons, and ionized air.

Taking $\rho \sim 1.3$ kg m^{-3} and the factor of 583.5 in (2.130), we find $K \sim 2.95 \times 10^{13}$ J. Explosion of one kiloton (kt) of TNT liberates 4.2×10^{12} J, so we have $K \sim 7.0$ kt. On accounting for the factor of two described above, our estimate of the total yield comes in at ~14 kt. The yield of the *Trinity* test is officially estimated as 21 kt, so this estimate is low by a factor of about one-third, but also recall from Sect. 2.6 that ~30% of the yield was due to fissions of ^{238}U, which has not been accounted for here. On considering the approximations involved, this is not at all a bad result for a quick calculation.

With the energy of the fireball in hand, a good student exercise would be to compute the time-evolution of pressure and temperature within the fireball, treating the enclosed air as an ideal gas; this would help drive home the phenomenal orders of magnitude of physical quantities involved in such explosions.

2.8.4 A Simplified Model of Tamped-Core Yield

The numerical simulation described in Sect. 2.6 tracks conditions within an exploding bomb core, as least as far as its underlying assumptions and the diffusion theory on which it is based are valid. Its disadvantage, however, is that it requires writing a program likely to run to a few hundred lines, which may make it impractical for a classroom setting. In this section, an approximate approach to estimating the yield of a tamped-core bomb is developed. What is lost in this approach is any detailed time-dependent tracking of various conditions, but it can be programmed into a single spreadsheet. Like the numerical simulation, this method is formulated so that the core and tamper experience the same radial expansion in order to model a uniform expansion velocity. Input densities can easily be altered to simulate any initial compression. This method is adopted from Reed (2018b).

The approach taken here is to treat the explosion in terms of the geometric growth of the neutron population, in combination with the already familiar idea that if the core starts with initial radius R_{init}, the explosion will proceed until second criticality occurs. At this point, the core radius reaches its "shutdown" value R_{core}^{shut}. The radial distance through which the core (and tamper) will have expanded is then $\Delta R = R_{core}^{shut} - R_{init}$. At the moment of shutdown, the core/tamper assembly will be just threshold critical, and so must satisfy Eq. (2.46), with R_{thresh} replaced by R_{core}^{shut} and R_{tamp} by the tamper radius at shutdown, $R_{tamp}^{shut} = R_{tamp}^{init} + \Delta R$:

$$\left[1 + \frac{2 R_{core}^{shut} \lambda_{tamp}^{trans}}{3 \left(R_{tamp}^{shut} \right)^2} - \frac{R_{core}^{shut}}{R_{tamp}^{shut}} \right] \left[\left(\frac{R_{core}^{shut}}{d_{core}} \right) \cot \left(\frac{R_{core}^{shut}}{d_{core}} \right) - 1 \right] + \frac{\lambda_{trans}^{tamp}}{\lambda_{trans}^{core}} = 0. \tag{2.134}$$

Let the core and tamper have masses M_{core} and M_{tamp}. Designating the densities of the core and tamper at any time by ρ_{core} and ρ_{tamp}, then their radii will be

$$R_{core} = \left(\frac{3 M_{core}}{4\pi \rho_{core}} \right)^{1/3} \tag{2.135}$$

and

$$R_{tamp} = \left[\frac{3}{4\pi} \left(\frac{M_{tamp}}{\rho_{tamp}} + \frac{M_{core}}{\rho_{core}} \right) \right]^{1/3}. \tag{2.136}$$

The premise of the calculation is, by trial and error, to find the value of ΔR which renders (2.134) satisfied, taking into account the fact that with each trial value of

ΔR, the densities of the core and tamper will change, and hence the values of d_{core} and the λ's in (2.134) must be modified accordingly. Otherwise, d_{core} and the λ's are as in Sects. 2.2 and 2.3. But for some notational differences, this is essentially the method that was followed in setting up Figs. 2.9 and 2.10. What is new here is the yield calculation that is about to be described. A significant player in this part of the model is the neutron travel time to fission:

$$\tau = \frac{\lambda_{fiss}^{core}}{v_{neut}}, \tag{2.137}$$

where v_{neut} is the average neutron speed.

To formulate the yield part of this model, I treat the fission chain reaction as a geometric-growth phenomenon. As in Sect. 2.5, suppose that N_O "initiator" neutrons are available to start the reaction. We saw there that by the end of G generations, the total number of fissions that will have occurred is given by

$$N_G = N_O \left(\frac{v^G}{v-1} \right). \tag{2.138}$$

If E_f is the energy liberated per fission, then the total energy liberated after G generations will be $E_f \, N_G$. With τ being the average time for a single fission generation, the G generations will correspond to elapsed time $t = G\tau$. As with the numerical simulation approach, I assume that all of the energy liberated to time t appears as the kinetic energy $M_{tot} v_{core}^2 / 2$ of the expanding (core + tamper) of total mass $M_{tot} = M_{core} + M_{tamp}$. Assuming that the (core + tamper) is expanding at a uniform rate throughout, the square of the expansion speed will be $(dr/dt)^2$, or

$$\left(\frac{dr}{dt} \right)^2 \sim \left[\frac{2 E_f N_O}{M_{tot}(v-1)} \right] e^{(\ln v / \tau) t}, \tag{2.139}$$

where $v^G = v^{(t/\tau)}$ has been written as $\exp(\ln v / \tau) \, t$. With R_{init} and R_{shut} as the initial and second-criticality radii of the core (the latter at time t_{shut}), we can solve Eq. (2.139) for dr/dt and integrate to give

$$\int_{R_{init}}^{R_{shut}} dr \sim \sqrt{\frac{2 E_f N_O}{M_{tot}(v-1)}} \int_0^{t_{shut}} e^{(\ln v / 2\tau) t} dt. \tag{2.140}$$

Carrying out the integral and setting $R_{core}^{shut} - R_{init} = \Delta R$ gives

$$t_{shut} \sim \left(\frac{2\tau}{\ln v} \right) \ln \left\{ (\Delta R) \left(\frac{\ln v}{2\tau} \right) \sqrt{\frac{M_{tot}(v-1)}{2 E_f N_O}} \right\}, \tag{2.141}$$

where the lower limit of integration of the time integral was dropped on the rationale that it will be small compared to the upper limit. Setting $G_{shut} = t_{shut}/\tau$ and computing the energy $E_f N_G$ liberated to shutdown with the help of (2.138) gives the yield as

$$E_{shut} \sim \left(\frac{E_f N_O}{\nu - 1}\right) \nu^{(t_{shut}/\tau)} \sim \frac{M_{tot}}{8} \left(\frac{\Delta R \ \ln \nu}{\tau}\right)^2, \qquad (2.142)$$

where the same exponential manipulation following (2.139) has been used. As with the approach of Sect. 2.5, the initial number of neutrons cancels out. This result possesses the same dependence of the yield on M_{core} and τ as does the more sophisticated analysis which led to (2.95). The dependence on ΔR is different, however: Calculations with the **CriticalityAnalytic** spreadsheet for untamped cores of about one to two critical masses show that $< \rho r >$, α, and ΔR are roughly linearly proportional to each other, which gives the numerator in (2.95) an overall dependence on ΔR as being roughly proportional to $(\Delta R)^4$. But this comparison should not be stretched too far: (2.95) applies to an untamped core, while the model in this section is for a tamped core. In (2.95), there is no explicit dependence on ν, which is implicit in α.

The question remains of what value of τ to use in (2.142). In the numerical simulation program, τ was adjusted at each timestep according as the core density. In the results related below, I use the value of τ at the time of second criticality on the rationale that the vast majority of the energy liberation occurs in the last few fission generations.

For the *Little Boy* model of Sect. 2.6 with a 53 kg core and 550 kg WC tamper, this model predicts a yield of 29.6 kt. Because the same threshold constraint was imposed in both models, they should both give the same expansion distance ΔR, and they do (2.30 cm). That the yield predictions are different is to be expected as they take very different approaches to calculating that quantity; for example, in the present model, there is no gas/radiation pressure constant γ. In this respect, time-dependent simulation surely does a better job of encoding the underlying physics, so we should probably not be too alarmed by the high yield indicated here. Again, predicted yields are very sensitive to input parameters: For the present model, leaving the core mass at 53 kg but reducing ν to 2.3 is by itself sufficient to reduce the predicted yield to ~14 kt. For *Fat Man*, taking a 6.3 kg core and 230 kg DU tamper with an initial compression ratio of 2.5, the yield emerges as 27 kt, again high compared with the numerical simulation, but not outrageously so.

Spreadsheets are available for performing the calculations involved in this "toy" model: **ToyYield(LB).xls** and **ToyYield(FM).xls**.

Figures 2.22 and 2.23 show some results from this model for uncompressed cores of ^{235}U with WC tampers. The curves in Fig. 2.22 show the expansion distances ΔR to second criticality for given core/tamper mass combinations. For example, a 56 kg core plus 600 kg tamper will expand by ~2.5 cm before criticality shutdown. Figure 2.23 shows corresponding yields; for this example, $Y \sim 35$ kt. The numerical simulation of Sect. 2.6 indicates $Y \sim 15.3$ kt for this arrangement. You should be able to verify the ~30 kt yield predicted for the 53 kg core/550 kg tamper combination described above.

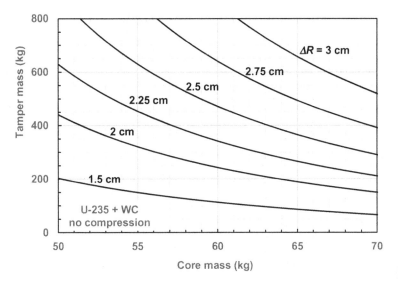

Fig. 2.22 Expansion distances to second criticality for given core/tamper mass combinations for a ^{235}U core and WC tamper, assuming no initial compression and same expansion distance for both core and tamper

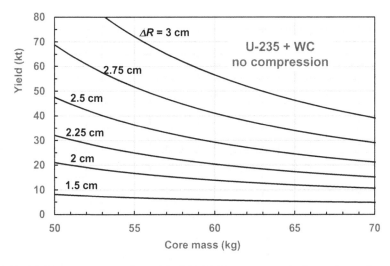

Fig. 2.23 Yield curves corresponding to Fig. 2.22. Yields are computed using total core plus tamper mass in (2.142)

2.9   Critical Mass of a Cylindrical Core (Optional)

The core of the *Little Boy* bomb was cylindrical in shape. However, all calculations so far have been predicated on spherical cores, so is natural to wonder how changing to a cylinder affects the calculation of critical mass.

It is difficult to analyze the situation for a cylindrical core because the boundary condition (2.29) that was used for the neutron diffusion equation in the spherical case,

$$N(R_C) = -\frac{2\,\lambda_t}{3}\left(\frac{\partial N}{\partial r}\right)_{R_C}, \tag{2.143}$$

is not easily generalized to the cylindrical case. However, if we are willing to admit a cruder boundary condition, much headway can be made with the cylindrical case. This is done in this section. This derivation can be considered optional as we consider only spherical cores in any subsequent section where the core geometry is relevant, such as in the analysis of predetonation in Chap. 4.

The cruder boundary condition is that the neutron density N is assumed to drop to zero at the surface for a cylinder of critical size. This situation is considered for a sphere in Sect. 2.8.1, and for a cube in Exercise 2.11. The critical volumes in those cases are

$$V_{sphere} = \left(\frac{4}{3}\pi^4\right)d^3 = 129.9\,d^3 \tag{2.144}$$

and

$$V_{cube} = \left(3^{3/2}\pi^3\right)d^3 = 161.1\,d^3, \tag{2.145}$$

where d is the characteristic length (2.25), which for threshold criticality ($\alpha = 0$) has the form

$$d = \sqrt{\frac{\lambda_f\,\lambda_t}{3\,(\nu - 1)}}. \tag{2.146}$$

For ^{235}U, d is about 3.5 cm.

Before beginning the formal solution, a few remarks on the diffusion equation in cylindrical coordinates are appropriate. Reactor engineers have been dealing with neutron fluxes in cylindrical geometries for decades, so the mathematics here, which involves *Bessel functions*, is not new. Bessel functions show up in a number of areas of mathematical physics such as quantum mechanics (the infinite cylindrical quantum well), acoustics (vibrations of drumheads), optics (diffraction through circular apertures) and electromagnetism (waveguides). Their appearance in criticality calculations illustrates connections between very different areas of physics.

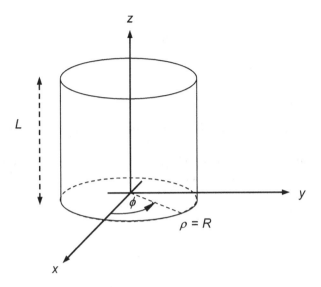

Fig. 2.24 Cylindrical core of radius R and height L

We begin with the general neutron diffusion equation of Appendix G:

$$\frac{\partial N}{\partial t} = \frac{v_{neut}}{\lambda_f}(\nu - 1)\,N + \frac{\lambda_t v_{neut}}{3}\left(\nabla^2 N\right). \tag{2.147}$$

The goal here is to apply this to the neutron population within a cylinder of radius R and length L as illustrated in Fig. 2.24. The bottom of the cylinder is imagined to be lying in the xy plane, with its center at $(x, y) = (0,0)$.

The separation of the diffusion equation into time and space-dependent parts proceeds as in Sect. 2.2; the temporal dependence is not of interest to us here as we seek to determine the threshold-critical condition. The spatial part of the neutron density N will be a function of the cylindrical coordinates (ρ, ϕ, z), and is assumed to be separable as

$$N_{\rho\phi z}(\rho, \phi, z) = N_\rho(\rho)N_\phi(\phi)N_z(z). \tag{2.148}$$

The Laplacian operator in cylindrical coordinates is

$$\nabla^2 N_{\rho\phi z} = \frac{1}{\rho}\frac{\partial}{\partial\rho}\left(\rho\frac{\partial N_{\rho\phi z}}{\partial\rho}\right) + \frac{1}{\rho^2}\frac{\partial^2 N_{\rho\phi z}}{\partial\phi^2} + \frac{\partial^2 N_{\rho\phi z}}{\partial z^2}. \tag{2.149}$$

On substituting (2.148) and (2.149) into (2.147) and dividing through by $N_{\rho\phi z}$, the spatial part of the diffusion equation appears, in analogy to (2.24), as

$$\frac{1}{d^2} + \frac{1}{N_\rho \rho} \frac{\partial}{\partial \rho} \left(\rho \frac{\partial N_\rho}{\partial \rho} \right) + \frac{1}{\rho^2 N_\phi} \frac{\partial^2 N_\phi}{\partial \phi^2} + \frac{1}{N_z} \frac{\partial^2 N_z}{\partial z^2} = 0. \tag{2.150}$$

The solution of this equation proceeds as does that of any separated differential equation. First, take the z-term to the right side:

$$\frac{1}{d^2} + \frac{1}{N_\rho \rho} \frac{\partial}{\partial \rho} \left(\rho \frac{\partial N_\rho}{\partial \rho} \right) + \frac{1}{\rho^2 N_\phi} \frac{\partial^2 N_\phi}{\partial \phi^2} = -\frac{1}{N_z} \frac{\partial^2 N_z}{\partial z^2}. \tag{2.151}$$

Since z is independent of ρ and ϕ, (2.151) can be true only if both sides are equal to a constant. This separation constant is traditionally defined to be $+k_z^2$, that is,

$$\frac{1}{N_z} \frac{\partial^2 N_z}{\partial z^2} = -k_z^2. \tag{2.152}$$

The solution of this equation is

$$N_z(z) = A e^{\iota k_z z} + B e^{-\iota k_z z}, \tag{2.153}$$

a result to which we will return presently.

Return to the left side of (2.151) and equate it to $+k_z^2$. Then multiply through by ρ^2 to clear that factor from the denominator of the ϕ term, move the ϕ term to the right side, and move the resulting $k_z^2 \rho^2$ term to the left side to effect another level of separation:

$$\frac{\rho}{N_\rho} \frac{\partial}{\partial \rho} \left(\rho \frac{\partial N_\rho}{\partial \rho} \right) + \left(\frac{1}{d^2} - k_z^2 \right) \rho^2 = -\frac{1}{N_\phi} \frac{\partial^2 N_\phi}{\partial \phi^2}. \tag{2.154}$$

As with (2.151), this expression can only be true if both sides are equal to a constant, which can be written as $+k_\phi^2$. This renders the ϕ-dependence as

$$\frac{1}{N_\phi} \frac{\partial^2 N_\phi}{\partial \phi^2} = -k_\phi^2, \tag{2.155}$$

which has the solution

$$N_\phi(\phi) = C e^{\iota k_\phi \phi} + D e^{-\iota k_\phi \phi}. \tag{2.156}$$

Now return to the left side of (2.154). Equate it to k_ϕ^2, and expand the derivative. This gives the radial dependence of the neutron density as

$$\rho^2 \frac{\partial^2 N_\rho}{\partial \rho^2} + \rho \left(\frac{\partial N_\rho}{\partial \rho} \right) + \left[\left(\frac{1}{d^2} - k_z^2 \right) \rho^2 - k_\phi^2 \right] N_\rho = 0. \tag{2.157}$$

If we now define

$$\kappa^2 = \left(\frac{1}{d^2} - k_z^2\right) \tag{2.158}$$

and establish the dimensionless variable

$$x = \kappa\rho, \tag{2.159}$$

Equation (2.157) becomes

$$x^2\frac{\partial^2 N_x}{\partial x^2} + x\left(\frac{\partial N_x}{\partial x}\right) + \left(x^2 - k_\phi^2\right)N_x = 0. \tag{2.160}$$

Note that x here is *not* the Cartesian-coordinate x, it is just a variable.

Equation (2.160) is *Bessel's equation of argument x and order k_ϕ*. Solutions to this physically important differential equation can be found in any good textbook on mathematical physics. However, we will not need to examine the detailed solutions; our interest is in satisfying the boundary condition that the neutron density falls to zero at the surface of the cylinder, $N(\text{edge}) = 0$.

Consider first the z-direction. In (2.153), we must demand $N_z(0) = 0$ and $N_z(L) = 0$. The first of these requires that $A + B = 0$, or $B = -A$; this gives

$$N_z(z) = A\left(e^{ik_z z} - e^{-ik_z z}\right), \tag{2.161}$$

which is equivalent to

$$N_z(z) = 2\iota A \sin(k_z z). \tag{2.162}$$

Now consider the condition $N_z(L) = 0$ applied to this result. This requires $\sin(k_z L) = 0$, which can only be satisfied if $k_z L$ is equal to an integer times π:

$$\sin(k_z L) = 0 \Rightarrow k_z = \frac{n\pi}{L}. \tag{2.163}$$

Now consider the ϕ-direction, where we have (2.156):

$$N_\phi(\phi) = Ce^{\iota k_\phi \phi} + De^{-\iota k_\phi \phi}. \tag{2.164}$$

Since there is no "edge" to the cylinder in the ϕ-direction, it is not immediately obvious what we should do with this expression. But the separation constant k_ϕ does appear in the radial Eq. (2.160), so we do need to pin it down somehow.

The condition to be applied to N_ϕ arises from the fact that ϕ is a so-called *cyclic coordinate*: If the value of ϕ is changed by adding any integral multiple of 2π radians, then one has returned to the same direction from which one began. We can express this by demanding that

$$N_\phi(\phi) = N_\phi(\phi + 2\pi),\tag{2.165}$$

or, more explicitly,

$$Ce^{\imath k_\phi \phi} + De^{-\imath k_\phi \phi} = Ce^{\imath k_\phi(\phi+2\pi)} + De^{-\imath k_\phi(\phi+2\pi)}.\tag{2.166}$$

This can be rewritten as

$$Ce^{\imath k_\phi \phi} + De^{-\imath k_\phi \phi} = Ce^{\imath k_\phi \phi}\left(e^{2\pi \imath k_\phi}\right) + De^{-\imath k_\phi \phi}\left(e^{-2\pi \imath k_\phi}\right).\tag{2.167}$$

This can *only* be satisfied if $e^{\pm 2\pi \imath k_\phi} = 1$, that is, if

$$\cos\left(2\pi k_\phi\right) \pm \imath \sin\left(2\pi k_\phi\right) = 1.\tag{2.168}$$

This expression will only be satisfied if

$$k_\phi = 0, 1, 2, 3, \ldots.\tag{2.169}$$

That ϕ is cyclic has led to the restriction that the *order* of our Bessel equation must be an integer.

With k_ϕ now established (at least to some extent), we can begin to get to the issue of the length and radius of a threshold-critical core. Return to the radial equation, (2.160):

$$x^2 \frac{\partial^2 N_x}{\partial x^2} + x\left(\frac{\partial N_x}{\partial x}\right) + \left(x^2 - k_\phi^2\right)N_x = 0.\tag{2.170}$$

The length L of the core appears explicitly in k_z, which is incorporated into this expression through κ and x.

To determine when criticality is achieved, we need to know what value(s) of x will just render (2.170) satisfied for a given value of the order k_ϕ; this will dictate the critical radius ρ through (2.159). For a given choice of k_ϕ, there prove to be an infinitude of values of x that make this so; these values are known as the *zeros of Bessel's equation for order k_ϕ*, and are extensively tabulated in many sources. In general, the values of the zeros increase monotonically within a given order, and the value of the m'th zero ($m = 1, 2, 3, \ldots$) also increases monotonically as a function of order number. The m'th zero for some order k is commonly designated as J_{km}; order numbers start at $k = 0$. In general, then, we will have criticality when x is equal to some zero J_{km}, or, on combining (2.158), (2.159), and (2.163), when

$$\left(\frac{1}{d^2} - \frac{n^2\pi^2}{L^2}\right)^{1/2} R = J_{km},\tag{2.171}$$

where the radius ρ is now written as R. The volume of the core is $\pi R^2 L$. We can solve (2.171) for R and then express the volume entirely in terms of L:

$$V_{crit} = \frac{\pi J_{km}^2 d^2 L^3}{\left(L^2 - n^2 \pi^2 d^2\right)}. \tag{2.172}$$

The lowest possible critical volume will obtain for the lowest possible value of J_{km} and the lowest possible value for n; we can choose these independently of each other as they arose from different separation constants. As for n, the lowest acceptable value is $n = 1$; $n = 0$ would not do as it would render $N_z(z) = 0$ *everywhere* throughout the core, not just at its edge [see (2.161) and (2.162)]. The lowest-valued zero J_{km} is $J_{01} = 2.40483$, that is, the first zero for the Bessel equation of order zero. This corresponds to $k_\phi = 0$, which is physically acceptable as it renders N_ϕ equal to a constant [see (2.164)]. The minimum critical volume then becomes

$$V_{crit} = \frac{\pi J_{01}^2 d^2 L^3}{\left(L^2 - \pi^2 d^2\right)}. \tag{2.173}$$

An interesting physical consequence here is that there is a minimum length required for the denominator to be positively-valued:

$$L > \pi d. \tag{2.174}$$

This result is intuitively appealing on the rationale that if the core is not long enough, too many neutrons will escape and criticality cannot be obtained. For ^{235}U, this critical length evaluates to about 11.04 cm.

The least possible critical volume is found by determining the value of L that minimizes (2.173). This proves to be

$$\frac{\partial V_{crit}}{\partial L} = 0 \Rightarrow L = \sqrt{3}\pi d, \tag{2.175}$$

which, when back-substituted into (2.173) gives

$$V_{min} = \left(\frac{3^{3/2}}{2}\pi^2 J_{01}^2\right) d^3 = 148.3 \, d^3. \tag{2.176}$$

For ^{235}U, this corresponds to a mass of about 121 kg. This result lies between those quoted at the beginning of this section for a sphere and a cube. The ratios of the critical volumes go as

$$V_{sphere} : V_{cyl} : V_{cube} = 1 : 1.142 : 1.241. \tag{2.177}$$

The penalty for using a *Little Boy*-type core instead of a sphere is thus only about a 14% increase in mass.

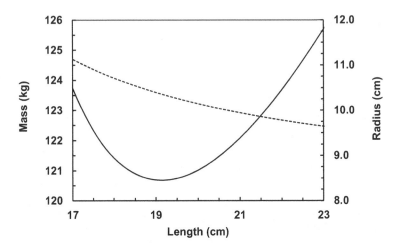

Fig. 2.25 Computed ^{235}U critical mass (solid line, left scale) and radius (dashed line, right scale) as a function of length for a cylindrical core. Note that these results hold only for the simplified boundary condition $N(\text{edge}) = 0$

Figure 2.25 shows the critical mass and cylinder radius corresponding to a given choice of L in (2.171) and (2.173) for our usual parameters for ^{235}U: $(\sigma_f, \sigma_{el}, \nu, \rho) = (1.235 \text{ bn}, 4.566 \text{ bn}, 2.637, 18.71 \text{ gr cm}^{-3})$. The minimum critical mass corresponds to a length of about 19.2 cm, and a radius of about 10.3 cm: A cylinder almost as long as it is wide.

References

Bernstein, J.: Hitler's Uranium Club: The Secret Recordings at Farm Hall. Copernicus Books, New York (2001)

Bernstein, J.: Heisenberg and the critical mass. Am. J. Phys. **70**(9), 911–916 (2002)

Bernstein. J.: A memorandum that changed the world. Am. J. Phys. **79**(5), 440–446 (2011)

Glasstone, S., Dolan, P.J.: The Effects of Nuclear Weapons. United States Department of Defense, Washington, DC (1977)

Koch, L.J., Paxton, H.C.: Fast reactors. Ann. Rev. Nuc. Sci. **9**, 437–472 (1959)

Logan, J.: The critical mass. Am. Sci. **84**, 263–277 (1996)

Malik, J.: The Yields of the Hiroshima and Nagasaki Nuclear Explosions, Los Alamos National Laboratory Report LA-8819 (1985). http://atomicarchive.com/Docs/pdfs/00313791.pdf

Peierls, R.: Critical conditions in neutron multiplication. Proc. Cam. Phil. Soc. **35**, 610–615 (1939)

Penney, W., Samuels, D.E.J., Scorgie, G.C.: The nuclear explosive yields at Hiroshima and Nagasaki. Phil. Trans. Roy. Soc. London. **A266**(1177), 357–424 (1970)

Reed, B.C.: Arthur Compton's 1941 Report on explosive fission of U-235: a look at the physics. Am. J. Phys. **75**(12), 1065–1072 (2007)

Reed, B.C.: Rudolf Peierls' 1939 analysis of critical conditions in neutron multiplication. Phys. Soc. **37**(4), 10–11 (2008)

Reed, B.C.: A brief primer on tamped fission-bomb cores. Am. J. Phys. **77**(8), 730–733 (2009)

Reed, B.C.: Student-level numerical simulation of conditions inside an exploding fission-bomb core. Nat. Sci. **2**(3), 139–144 (2010)

Reed, B.C.: Note on the minimum critical mass for a tamped fission bomb core. Am. J. Phys. **83**(11), 969–971 (2015)

Reed, B.C.: A physicists guide to The Los Alamos Primer. Phys. Scr. **91**(11), 113002 (30 pp) (2016). (Erratum: Phys. Scr. **91**(12) 129601 (1p), 2016)

Reed, B.C.: A simple model for the critical mass of a nuclear weapon, Phys. Educ. **53**, 043002 (3 pp) (2018a)

Reed, B.C.: A toy model for the yield of a tamped fission bomb. Am. J. Phys. **86**(2), 105–109 (2018b)

Reed, B.C.: Composite cores and tamper yield: lesser-known aspects of Manhattan Project fission bombs. Am. J. Phys. **88**(2), 108–114 (2020a)

Reed, B.C.: Estimating the yield of the *Trinity* test with a simple kinetic energy argument. Phys. Educ. **55**, 033007 (3 pp) (2020b)

Reed. B.C.: A powerful graphical display of technical information: Robert Serber's plot of physical conditions inside a nuclear explosion. Am. J. Phys. **88**(7) 565-567 (2020c)

Semkow, T.M., Parekh, P.P., Haines, D.K.: Modeling the effects of the trinity test. In Applied Modeling and Computations in Nuclear Science. American Chemical Society Symposium Series, vol. 945, pp. 142–159 (2006)

Serber, R.: The Los Alamos Primer: The First Lectures on How To Build An Atomic Bomb. University of California Press, Berkeley (1992)

Soodak, H. (ed.): Reactor Handbook, 2nd edn. Part A: Physics, vol. 3. Interscience Publishers, New York (1962)

Taylor, G.: The formation of a blast wave by a very intense explosion I. Theoretical discussion. Proc. R. Soc. A **201**(1065), 159–174 (1950a)

Taylor, G.: The formation of a blast wave by a very intense explosion II. The atomic explosion of 1945. Proc. R. Soc. A **201**(1065) 175–186 (1950b)

Chapter 3
Producing Fissile Material

The vast majority of the manpower and funding involved in the Manhattan Project were devoted to producing fissile material. Uranium-235 had to be separated from natural uranium, and plutonium had to be synthesized in nuclear reactors. In this chapter we examine some of the physics behind these processes. Historically, the first major step along these lines occurred when Enrico Fermi and his collaborators achieved the first operation of a self-sustaining chain-reaction on December 2, 1942, with their CP-1 ("Critical Pile 1" or "Chicago Pile 1") reactor. This proved that a chain-reaction could be created and controlled, and opened the door to the design and development of large-scale plutonium-production reactors located at Hanford, WA. We thus look first at issues of reactor criticality (Sects. 3.1 and 3.2), and then examine plutonium production (Sect. 3.3). Sections 3.4 and 3.5 are devoted to analyzing techniques for enriching uranium.

3.1 Reactor Criticality

The key quantifier in achieving a self-sustaining chain reaction is what is known as the "criticality factor" or "reproduction factor," designated as k. This dimensionless number is defined in such a way that if $k \geq 1$, then a reaction will be self-sustaining, whereas if $k < 1$, the reaction will eventually die out. In fact, if $k > 1$, the reaction rate will grow exponentially; reactors are equipped with control mechanisms that can be adjusted to maintain $k = 1$. k is analogous to the secondary neutron number ν that was involved in the discussion of critical mass and efficiency in the preceding chapter. k is also known as the "reproduction constant", although this terminology is somewhat of a misnomer.

Achieving a chain reaction with uranium of natural isotopic composition involves several competing factors. The small fraction of ^{235}U present is inherently extremely fissile when bombarded by slow neutrons, and, for each neutron consumed in

© The Author(s), under exclusive license to Springer Nature Switzerland AG 2021   119
B. C. Reed, *The Physics of the Manhattan Project*,
https://doi.org/10.1007/978-3-030-61373-0_3

fissioning a ^{235}U nucleus, some 2.4 are on average released; these can go on to initiate other fissions. On the other hand, the vastly more abundant ^{238}U nuclei tend to capture neutrons without fissioning, removing them from circulation.

When a nucleus is struck by a neutron, one of three things will in general happen: (i) The nucleus may fission; (ii) The nucleus may capture the neutron without fissioning; or (iii) The neutron may scatter from the nucleus. This last process serves only to redirect neutrons within the reactor and can be ignored if the reactor is sufficiently large that a neutron has a good chance of being involved in a fission or capture before being scattered through the surface of the reactor and lost. We will be concerned with processes (i) and (ii).

The likelihood of each process is quantified by a corresponding cross-section. We will be concerned with fission (f) and capture (c) cross-sections for ^{235}U and ^{238}U. In self-evident notation, we write these as σ_{f5}, σ_{c5}, σ_{f8}, and σ_{c8}. Numerical values for these quantities are listed in Table 3.1 for both isotopes for both "fast" and "slow" neutrons, also known as "unmoderated" and "moderated" neutrons, respectively. For the latter, the cross-sections refer to neutrons of kinetic energy 0.0253 eV; the origin of this number is explained in Sect. 3.2. Sources for these values are given in Appendix B. Also shown are the average numbers of secondary neutrons for each isotope for both fast- and slow-neutron induced fissions. Two important things to notice here are (i) The large fission cross-section for *slow* neutrons in the case of ^{235}U, and (ii) The non-zero *capture* cross-section for *slow* neutrons for the same isotope: Upon capturing a slow neutron, a ^{235}U nucleus has about a one-in-seven chance of *not* fissioning. Figure 3.1 shows the fission cross-section of ^{235}U from 10^{-8} MeV to 1 MeV.

The value given in Table 3.1 for the capture cross-section of ^{238}U for fast neutrons, 2.661 bn, is the sum of this isotope's *true* capture cross-section for fast neutrons (0.0664 bn) plus its inelastic scattering cross-section for fast neutrons (2.595 bn). The rationale for this is that when neutrons inelastically scatter from ^{238}U, they lose so much energy as to fall below the fission threshold for that isotope and are virtually guaranteed to be captured should they strike another ^{238}U nucleus; inelastic scattering by ^{238}U is therefore effectively equivalent to capture by it (see Sect. 1.9). We also assume that no neutrons are lost due to capture by fission products; in reality, this is not a trivial issue.

Table 3.1 Fissility parameters

Parameter	Fast neutrons	Slow neutrons
σ_{f5} (bn)	1.235	584.4
σ_{c5} (bn)	0.08907	98.81
ν_5	2.637	2.421
σ_{f8} (bn)	0.3084	0
σ_{c8} (bn)	2.661	2.717
ν_8	2.655	2.448

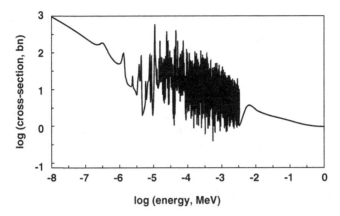

Fig. 3.1 Fission cross-section for ^{235}U as a function of neutron energy in MeV, from 10^{-8} MeV to 1 MeV; Data from Korean Atomic Energy Research Institute file pendfb7/U235:19. For thermal neutrons (log $E = -7.6$), the cross-section is 585 barns. Only about 5% of the available data is plotted here. Many of the resonance capture spikes are so finely spaced that they cannot be resolved

Suppose that our reactor consists of a mixture of ^{235}U and ^{238}U isotopes. (A more advanced version of this calculation which takes into account the neutron-capturing effects of graphite and any impurities appears in Sect. 4.1; the present approach is the simplest—and hence most optimistic—version). For each isotope, we write a total cross section as the sum of the cross-sections for all individual processes involving that isotope:

$$\sigma_5 = \left(\sigma_{f5} + \sigma_{c5}\right), \tag{3.1}$$

and

$$\sigma_8 = \left(\sigma_{f8} + \sigma_{c8}\right). \tag{3.2}$$

Let the fractional abundance of ^{235}U be designated by F; $0 \leq F \leq 1$. For neutrons created in fissions or otherwise supplied, the total cross-section for them to suffer *some* subsequent process is given by the abundance-weighted sum of the cross-sections for the individual isotopes:

$$\sigma_{total} = F\,\sigma_5 + (1 - F)\,\sigma_8. \tag{3.3}$$

Now imagine following a single neutron as it flies about within the reactor until it causes a fission or is captured. The reproduction factor k is defined as the average number of neutrons that this one original neutron subsequently gives rise to. We derive an expression for k by separately computing the number of secondary neutrons created by fissions of ^{235}U and ^{238}U, and then adding the two results.

The probability of our neutron striking a nucleus of ^{235}U is given by the ratio of the total cross-section for ^{235}U to that for the total cross-section for all processes and isotopes, weighted by the abundance fraction of that isotope: $(F\sigma_5/\sigma_{total})$. Once the neutron has struck the ^{235}U nucleus, the probability that it initiates a fission will be (σ_{f5}/σ_5). Hence, by the usual multiplicative process for combining independent probabilities, the overall probability that the neutron will fission a ^{235}U nucleus will be the product of these factors, $(F\sigma_5/\sigma_{total})(\sigma_{f5}/\sigma_5) = (F\sigma_{f5}/\sigma_{total})$. If fission of a ^{235}U nucleus liberates on average ν_5 secondary neutrons, then the average number of neutrons created by one neutron from fissioning a ^{235}U nucleus will be $\nu_5(F\sigma_{f5}/\sigma_{total})$. Likewise, fissions of ^{238}U nuclei will give rise, on average, to $\nu_8(1-F)(\sigma_{f8}/\sigma_{total})$ secondary neutrons. The total number of secondary neutrons created by one initial neutron will then be

$$k = \frac{F\,\nu_5\,\sigma_{f5} + (1-F)\nu_8\,\sigma_{f8}}{\sigma_{total}}. \tag{3.4}$$

In natural uranium, $F = 0.0072$; we ignore here the very small natural abundance of ^{234}U.

We first apply (3.4) to *unmoderated* (fast) neutrons and natural uranium:

$$\sigma_{total} = F\left(\sigma_{f5} + \sigma_{c5}\right) + (1-F)\left(\sigma_{f8} + \sigma_{c8}\right)$$
$$= (0.0072)(1.32407) + (0.9928)(2.9694) = 2.958 \text{ bn.} \tag{3.5}$$

Hence

$$k_{fast} = \frac{F\,\nu_5\,\sigma_{f5} + (1-F)\nu_8\,\sigma_{f8}}{\sigma_{total}}$$
$$= \frac{(0.0072)(2.637)(1.235) + (0.9928)(2.655)(0.3084)}{2.958}$$
$$= 0.283. \tag{3.6}$$

Since $k_{fast} < 1$, *a self-sustaining chain reaction using unmoderated neutrons with natural uranium is impossible*. This is why a lump of ordinary uranium of any size is perfectly safe against a spontaneous chain reaction; a nuclear weapon cannot be constructed using uranium of natural isotopic composition.

Figure 3.2 shows k_{fast} as a function of F; k_{fast} does not exceed unity until $F \sim 0.53$. One must undertake a significant enrichment effort to construct a uranium bomb. Bomb-grade uranium is usually considered to be 90% ^{235}U ($k \sim 2.02$).

In the case of moderated neutrons, the story is very different. Here we have

$$\sigma_{total} = (0.0072)(584.4 + 98.81) + (0.9928)(0 + 2.717) = 7.617 \text{ bn.} \tag{3.7}$$

Fig. 3.2 Fast-neutron reproduction factor k versus ^{235}U abundance fraction

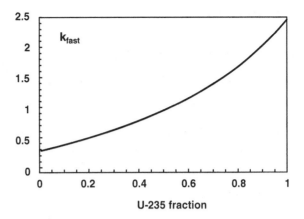

U-235 fraction

Hence

$$k_{slow} = \frac{F\, v_5\, \sigma_{f5} + (1 - F)v_8\, \sigma_{f8}}{\sigma_{total}}$$

$$= \frac{(0.0072)\,(2.421)\,(584.4) + 0}{7.617} = 1.337. \tag{3.8}$$

Since $k > 1$, a self-sustaining reaction with moderated neutrons and natural uranium is possible. This slow-neutron reproduction factor is the premise underlying CP-1 and all commercial power-producing reactors. In situations where a commercial-scale reactor would be impractical due to its physical size (such as in a naval vessel), smaller reactors fueled with uranium significantly enriched in ^{235}U are used.

If ^{235}U has such an enormous fission cross-section for slow neutrons, why not build a bomb that incorporates a moderator and utilizes slow neutrons? The reason for this can be seen in the efficiency formula of Eq. (2.96): the energy liberated by a nuclear weapon is proportional to the inverse-square of the time required for a neutron to travel from where it is born in a fission to where it causes a subsequent fission. This time is itself inversely proportional to the speed of a neutron, so the time squared will be inversely proportional to the square of the speed of the neutron, that is, to its kinetic energy. Hence, a bomb utilizing slow neutrons with kinetic energies ~0.025 eV would liberate only about 10^{-8} times as much energy as that released by one which utilizes fast neutrons with kinetic energies of ~2 MeV. If a fast-neutron bomb is designed to explode with an energy of 20 kilotons TNT equivalent, a "corresponding" slow-neutron bomb would release less energy than a single pound of TNT! There is simply no point in making a slow-neutron bomb; in effect, one might as well attempt to drop a reactor on an adversary.

3.2   Neutron Thermalization

Fermi's CP-1 reactor used graphite (crystallized carbon) as a *moderator* to slow neutrons emitted by fissioning ^{235}U nuclei to so-called "thermal" speeds to take advantage of the large fission cross-section of that isotope for neutrons of such energy. Graphite was used as it has a small capture cross-section for neutrons. In this section we quantify the meaning of "thermal," and estimate the typical distance a neutron will travel during the thermalization process; this will give us insight as to why the lumps of uranium in CP-1 were distributed as a cubical lattice with a spacing of 8.25 in. (21 cm). A detailed description of CP-1 was published in Fermi (1952).

To quantify what is meant by a thermal neutron, recall from Maxwellian statistical mechanics that the most probable velocity of a particle of mass m at absolute temperature T is given by

$$v_{mp} = \sqrt{\frac{2k_B T}{m}}, \tag{3.9}$$

where k_B is Boltzmann's constant. Thermalization is taken to correspond to $T = 298$ K (about 77 °F), that is, approximately room temperature. For neutrons, this evaluates to

$$v_{mp} = \sqrt{\frac{2\left(1.381 \times 10^{-23}\text{J/K}\right)(298\text{K})}{\left(1.675 \times 10^{-27}\text{kg}\right)}} = 2217\text{m/s}. \tag{3.10}$$

The kinetic energy of such a neutron is

$$E = \frac{1}{2}mv_{mp}^2 = 4.115 \times 10^{-21}\text{J} = 0.025\text{eV}. \tag{3.11}$$

The physical premise involved here is that since the nuclei of the moderating material will be randomly moving with energies characteristic of room temperature, neutrons cannot on average be slowed to lower speeds via collisions with the moderating material. More precisely, "thermal" neutrons are defined in technical nuclear physics literature to have $v = 2200$ m/s, which corresponds to an energy of 0.0253 eV. This value is much less than the typical ~2 MeV with which secondary neutrons emerge from a fissioned nucleus.

Nuclear physicists often use the concepts of "kinetic energy" and "temperature" interchangeably in the above sense; you should be able to show that if kinetic energy as computed using the most probable speed of (3.9) is expressed in units of eV, then the equivalent temperature in Kelvins is given approximately by T ~ 11,600 (KE). A temperature of a million Kelvins corresponds to a kinetic energy of about 86 eV. The center of the Sun is estimated to have a temperature of 15 million Kelvins, which corresponds to KE ~ 1300 eV. A fission fragment with a kinetic energy of 100 MeV thus has an equivalent temperature of just over a *trillion* Kelvins.

Fig. 3.3 Elastic collision between a neutron of mass m and a nucleus of mass M. The two are treated as non-deformable, non-rotating disks or spheres

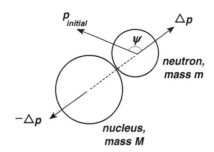

If graphite is used as the moderator, how many times will a neutron scatter until it becomes thermalized, and how far will it travel in doing so? Consider a neutron of mass m that strikes an initially stationary nucleus of mass M as sketched in Fig. 3.3. If the environment is thermal, it will not be a drastic approximation to treat the struck nucleus as stationary in comparison to a presumably much less massive neutron that has been emitted in a fission. The collision is treated as being elastic, with both the neutron and nucleus modeled as smooth, non-deformable, non-rotating disks (or spheres). During the collision, they exert equal and opposite impulses on each other, which are directed along the line joining their centers; if they are smooth, they can exert no tangential forces on each other. The neutron has initial momentum $\boldsymbol{p}_{initial}$, and suffers change in momentum $\Delta \boldsymbol{p}$; the nucleus acquires momentum $-\Delta \boldsymbol{p}$. The plane of the sketch is the plane containing both $\boldsymbol{p}_{initial}$ and $\Delta \boldsymbol{p}$.

The post-collision momentum of the neutron will be $\boldsymbol{p}_{initial} + \Delta \boldsymbol{p}$. Kinetic energy is given by $(\boldsymbol{p} \bullet \boldsymbol{p})/2m$, so the post-collision kinetic energy of the neutron will be

$$K_{final} = \frac{1}{2m} \left(\boldsymbol{p}_{initial} + \Delta \boldsymbol{p} \right) \bullet \left(\boldsymbol{p}_{initial} + \Delta \boldsymbol{p} \right). \tag{3.12}$$

Expanding this gives

$$K_{final} = K_{initial} + \frac{p_{initial} \left(\Delta p \right) \cos \psi}{m} + \frac{\left(\Delta p \right)^2}{2m}, \tag{3.13}$$

where ψ is the angle between the original direction of the neutron's motion and $\Delta \boldsymbol{p}$.

In an elastic collision, total system kinetic energy is conserved, so we must also have

$$\frac{\left(p_{initial} + \Delta_p \right) \cdot \left(p_{initial} + \Delta_p \right)}{m} + \frac{\left(-\Delta_p \right) \cdot \left(-\Delta_p \right)}{M} = \frac{\left(p_{initial} \right) \cdot \left(p_{initial} \right)}{m}.$$

Expanding out this expression and simplifying gives

$$\Delta p = -\left(\frac{2M}{m + M}\right) p_{initial} \cos \psi. \qquad (3.14)$$

Substituting (3.14) back into (3.13) gives, after some algebra, a very compact expression for the ratio of the post-collision to pre-collision kinetic energies of the neutron:

$$\frac{K_{final}}{K_{initial}} = 1 - \beta \cos^2 \psi, \qquad (3.15)$$

where

$$\beta = \frac{4mM}{(m + M)^2}. \qquad (3.16)$$

By sketching a few collisions of the form of Fig. 3.3, you should be able to convince yourself that $\pi/2 \leq \psi \leq \pi$, always. For heavy target nuclei, β will be small, and the neutron can lose anywhere from very little to none of its kinetic energy in one collision. For N successive collisions, the final kinetic energy will be given by the product of N terms of the form of (3.15), each with its own value of ψ. To get an *approximate* estimate of the number of collisions necessary to reach some final kinetic energy, I adopt the average value of $\cos^2 \psi$ over the range $(\pi/2 < \psi < \pi)$, which is $1/2$. Use this in (3.15)), and then take logarithms to give

$$N \sim \frac{\ln\left(K_{final}/K_{initial}\right)}{\ln\left(1 - \beta/2\right)}. \qquad (3.17)$$

In nuclear engineering literature, K_{final} is often taken to be 1 eV; choosing a lower value would get into the issue of not being able to ignore any initial momentum of the target nucleus.

Table 3.2 shows results for common moderating materials, taking $K_{initial} = 2\,\mathrm{MeV}$ and $K_{final} = 1$ eV. The last column of the table shows results for the same energy decrement adopted from a well-regarded nuclear power website. It is reassuring that the present approach gives results in reasonable accord with these, but this must be somewhat a matter of canceling errors in view of the approximations invoked.

As to the thermalization distance, refer to the derivation in Sect. 2.1 of the average distance a particle can be expected to penetrate through a medium before suffering a reaction. In application to the present case, we can write this as a mean free path as in Eq. (2.14):

$$\lambda_s = \frac{1}{\sigma_s n}, \qquad (3.18)$$

Table 3.2 Number of collisions necessary to thermalize neutrons of $K_{initial} = 2$ MeV to $K_{final} = 1$ eV

Element	Mass A	β	Collisions to thermalization (this work)	Collisions to thermalize (Website[a])
H	1	1	21	15
D	2	0.8888	25	20
He	4	0.64	38	34
Be	9	0.36	73	70
C	12	0.2840	95	92
O	16	0.2215	124	121

[a]https://www.nuclear-power.net/glossary/neutron-moderatoraverage-logarithmic-energy-decrement/

where σ_s is the scattering cross-section and n is the number density of nuclei. Strictly, this applies only for neutrons scattering through a medium of infinite extent, but since any sensible reactor will have a size considerably greater than λ_s, this is not a problem.

The density of graphite is 1.62 gr cm^{-3}, for which $n \sim 8.13 \times 10^{28}$ m^{-3}. For thermal neutrons, the elastic scattering cross-section for ^{12}C is 4.746 bn; this number is taken from the KAERI site referenced in Appendix B. These figures give

$$\lambda_s \sim 2.6 \text{ cm.} \tag{3.19}$$

This is equivalent to about one inch. Now, we know from statistical mechanics that if a particle takes N randomly-directed steps of length λ from some starting point, then the resulting average displacement from the starting point will be $\sqrt{N}\,\lambda$. In the present case, the neutron displacement will be $\sqrt{N}\,\lambda_s \sim \sqrt{95}(2.6$ cm$) \sim 25$ cm, a figure close to CP-1's 21 cm lattice spacing. Fermi designed CP-1 to occupy the minimum volume possible while achieving effective neutron thermalization.

3.3 Plutonium Production

The three giant graphite-moderated, water-cooled plutonium production piles constructed for the Manhattan Project in Hanford, Washington, were vastly scaled-up, much more complex versions of Fermi's CP-1 pile (Fig. 3.4). Fueled with natural uranium, these reactors were designed to utilize a controlled slow-neutron chain-reaction as described in the preceding two sections to synthesize ^{239}Pu from neutron capture by ^{238}U and subsequent beta-decay:

$$_{0}^{1}\text{n} + _{92}^{238}\text{U} \rightarrow _{92}^{239}\text{U} \xrightarrow[23.5 \text{ min}]{\beta^-} _{93}^{239}\text{Np} \xrightarrow[2.36 \text{ days}]{\beta^-} _{94}^{239}\text{Pu.}$$

Fig. 3.4 Workers laying the graphite core of a Hanford reactor. The rear face of the reactor is toward the lower left, and the inside of the front face to the upper right. Including shielding, the outer dimensions of each pile were 37 by 46 ft in footprint by 41 ft high. The graphite core for each pile measured 36 ft wide by 36 ft tall by 28 ft from front-to-rear. Each pile comprised some 75,000 graphite bricks about 4 in. square by 4 ft long, with one in every five bored lengthwise to accommodate fueling tubes spaced about 8 in. apart. Each reactor had 2004 fueling tubes; a full load of fuel comprised about 250 t of natural-uranium slugs about 9 in. long and 1.4 in. in diameter (Historic American Engineering Record 2000, Photo 6)

The important question is the rate of plutonium production in grams or kilograms per day. The answer to this can be gleaned from knowledge of the power output of the reactor, the isotopic composition of the fuel, and the fission and capture cross-sections for the isotopes involved. The analysis presented in this section is adopted from Reed (2005, 2019).

Commercial power-producing reactors are usually rated by their net electrical power output, so many "megawatts electrical," which is designate by the symbol P_e. However, this quantity reflects power output after accounting for inevitable thermal (Carnot) inefficiencies involved. The power output "within" the reactor itself—the number of "megawatts thermal"—is given by $P_t = P_e/\eta$, where η is the thermal efficiency of the plant. Typically, $\eta \sim 0.3$–0.4. In the case of a reactor, the power produced derives from mass-energy liberated in the fissioning of ^{235}U atoms. Various fission reactions are possible, but we can simplify the situation by assuming that each liberates, on average, energy E_f. The rate of reactions R within the pile must then be given by $R = P_t/E_f$, with P_t in Watts and E_f in Joules; conversion to more convenient units will be made presently.

In the calculation which follow, I assume that the creation of ^{239}Pu via neutron capture by ^{238}U is an essentially instantaneous process; no account is taken of the 23 min and 2.4 day half-lives of the intermediate ^{239}U and ^{239}Np nuclei, nor of any depletion of them by other reactions. So far as the time-scales go, this is quite reasonable: Fuel to produce Pu during the Manhattan Project was usually left in reactors for ~100 days.

To estimate plutonium production, it is helpful to introduce the idea of the reaction rate expressed in terms of the neutron flux within the reactor. This was touched on in Sect. 2.1. The neutron flux Φ is the number of neutrons passing through a unit area (such as 1 cm^2) per second. If the neutrons are incident on N target nuclei of reaction cross-section σ, then the reaction rate will be $R = \Phi N \sigma$ per second; see the development leading to Eq. (2.2). In a reactor, the number of nuclei involved in power production will be those of ^{235}U, and the relevant cross-section will be that for fission by thermal neutrons of that isotope, σ_{f5}. Hence we have

$$R = \frac{P_t}{E_f} = \Phi N_{235} \sigma_{f5} \quad \Rightarrow \quad \Phi = \frac{P_t}{E_f N_{235} \sigma_{f5}}. \tag{3.20}$$

To obtain the rate of synthesis of Pu nuclei, use the rate equation again, but with the number of nuclei of ^{238}U and its capture cross-section for thermal neutrons, σ_{c8}:

$$R_{Pu} = \Phi N_{238} \sigma_{c8} = \frac{P_t N_{238} \sigma_{c8}}{E_f N_{235} \sigma_{f5}}. \tag{3.21}$$

The ratio of the number of nuclei can be expressed in terms of the fractional abundance F of ^{235}U in the reactor's fuel:

$$R_{Pu} = \frac{P_t (1 - F) \sigma_{c8}}{E_f F \sigma_{f5}}. \tag{3.22}$$

Reactor powers are normally quoted in megawatts (MW; P_{MW}), and fission energies in MeV. Making these conversions, (3.22) becomes

$$R_{Pu} = \left(6.241 \times 10^{18}\right) \frac{P_{MW} (1 - F) \sigma_{c8}}{E_f^{MeV} F \sigma_{f5}}. \tag{3.23}$$

This expression gives the rate of production of Pu in nuclei per second. On accounting for the mass of ^{239}Pu nuclei ($239.05u = 3.970 \times 10^{-25}$ kg) and the number of seconds in a day, we can transform this into an expression for the number of grams of Pu produced per day:

$$R_{Pu} = 214.1 \left[\frac{P_{MW} (1 - F) \sigma_{c8}}{E_f^{MeV} F \sigma_{f5}} \right] \text{(gr day}^{-1}\text{)}. \tag{3.24}$$

For various reasons, power-producing reactors in the United States use fuel enriched to $F \sim 0.03$. For a plant producing electric power at a rate of 1 GW fueled at $F = 0.03$ and operating at efficiency $\eta = 0.3$ ($P_{MW} = 3333$), (3.24) gives a production rate of 593 gr day^{-1} = 216 kg yr^{-1}, assuming $\sigma_{f5} = 584$ b, $\sigma_{c8} = 2.7$ b, and E_f

$= 180$ MeV. Given that there are some 100 commercial reactors in operation in the United States, we can infer the annual production of plutonium to be on the order of 20,000 kg, enough for more than 2000 Nagasaki-type bombs. However, commercial-reactor fuel rods in the United States are not reprocessed, so the Pu created remains locked up in them. Ironically, ^{239}Pu α-decays back to ^{235}U with a half-life of about 24,000 years; our distant descendants will find a fresh supply of "enriched" fuel rods awaiting them! A reactor which produces one gigawatt of electricity at $\eta = 0.3$ but fueled with natural uranium ($F = 0.0072$, such as used in the Canadian CANDU system) will produce some 920 kg of Pu per year.

Fuel rods typically remain in commercial reactors for months or years. A result of this is that some of the ^{239}Pu nuclei that are formed have time to capture neutrons to become nuclei of ^{240}Pu. As we explore in Sect. 4.2, this isotope is characterized by an extremely high spontaneous fission rate, a situation which presents a dangerous challenge for anyone who seeks to construct a nuclear weapon from such spent fuel. An excellent treatment of issues in civilian nuclear power generation appears in Garwin and Charpak (2001).

For the present purposes, our interest is with the Hanford reactors, which were fueled with natural uranium and operated at a thermal power $P_t = 250$ MW. For these figures, (3.24) gives a production rate of about 0.76 grams per MW per day, or 190 gr day^{-1}. Three reactors operating at this power would produce 570 gr day^{-1}. To synthesize enough Pu to construct a bomb core of 6 kg would therefore require about 11 days of steady-state operation. Fuel slugs were left in the Hanford reactors for typically 100 days of neutron bombardment; after being withdrawn, they had to be cooled, and time allowed for dangerous short-lived fission products to decay. A discussion of the design of these reactors appears in Weinberg (2002).

The reaction rate Eq. (3.20), cast in terms of the mass of fuel in a reactor, will be used again in Sect. 4.5 to estimate the rate of production of polonium in a reactor via neutron bombardment of bismuth. Polonium was used in Manhattan Project bombs in neutron-emitting "initiators" used to trigger the nuclear chain reactions in the *Little Boy* and *Fat Man* bombs.

3.4 Electromagnetic Separation of Isotopes

The Manhattan Project's Oak Ridge, Tennessee, facility was devoted to enriching uranium for use in the *Little Boy* bomb. Three separate techniques were involved in this effort: (i) Electromagnetic separation; (ii) Gaseous (barrier) diffusion, and (iii) Liquid thermal diffusion. The first two of these can be examined on the basis of undergraduate-level physics, and are so treated in this and the following sections. The physics of liquid thermal diffusion is extremely complex, however, so we do not consider that process further. Readers interested in the technical details of liquid thermal diffusion are urged to consult the classic paper "The Separation of Isotopes by Thermal Diffusion" by Jones and Furry (1946). A qualitative description of the use of thermal diffusion in the Manhattan Project can be found in Reed (2011).

We first deal with electromagnetic separation of isotopes. Barrier diffusion is taken up in Sect. 3.5.

The electromagnetic separation facility at Oak Ridge was code-named Y-12, and utilized "calutron" separators designed by Ernest Lawrence; the name is a contraction of "California University Cyclotron." The design of these separators was predicated on the phenomenon that an ion, when directed into a magnetic field oriented perpendicularly to the its initial velocity, will subsequently travel in a circular orbit whose radius is dictated by the strength of the field, the magnitude of the initial velocity, the degree of ionization, and the ion's mass. Isotopes of different masses will consequently travel in different orbits, and can be separated. As with any isotope separation technique, this method depends on the very slight mass difference between the isotopes involved. In the case of ^{235}U and ^{238}U, the mass difference is very small, so this technique is extremely difficult to realize in practice.

To analyze this, we use a coordinate system where the x and y axes are in the plane of the page and the z-axis is directed out of the page as shown in Fig. 3.5.

Assume that a uniform magnetic field $\vec{B} = B\,\hat{z}$ emerges perpendicularly from the page. An ion of mass m and net charge q (usually positive) moves under the influence of the field. According to the Lorentz force law, the force on the ion at any time will be

$$\vec{F} = q\left(\vec{v} \times \vec{B}\right) = q\,B\left(v_y\hat{x} - v_x\hat{y}\right). \tag{3.25}$$

Newton's Second law holds that $\vec{F} = m\vec{a}$, so we can write

$$q\,B\left(v_y\hat{x} - v_x\hat{y}\right) = m\left(\frac{dv_x}{dt}\hat{x} + \frac{dv_y}{dt}\hat{y} + \frac{dv_z}{dt}\hat{z}\right), \tag{3.26}$$

from which we have

$$\frac{dv_x}{dt} = \alpha\,v_y \tag{3.27}$$

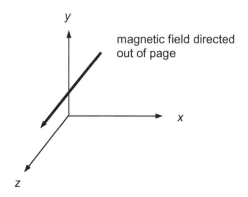

Fig. 3.5 Coordinate system for analyzing motion of charged particles in a magnetic field. The x and y axes are in the plane of the page; the z-axis emerges from the page, as does the magnetic field

magnetic field directed out of page

and

$$\frac{dv_y}{dt} = -\alpha\, v_x, \tag{3.28}$$

where

$$\alpha = \frac{q\,B}{m}. \tag{3.29}$$

The α here is a completely different quantity from the neutron exponential growth factor in Chap. 2. Equations (3.27) and (3.28) are *coupled* differential equations: The rate of change of v_x depends on v_y, and vice versa. Note that we must have $dv_z/dt = 0$; if the ion enters the magnetic field with $v_z = 0$, its subsequent motion will be restricted to the xy plane, the case assumed here.

Equations (3.27) and (3.28) can be separated by the following manipulation. Differentiate (3.27) with respect to time:

$$\frac{d^2 v_x}{dt^2} = \alpha\, \frac{dv_y}{dt}. \tag{3.30}$$

Now substitute (3.30) into (3.28) to eliminate dv_y/dt:

$$\frac{d^2 v_x}{dt^2} = -\alpha^2\, v_x. \tag{3.31}$$

What we have gained here is a differential equation that involves only the x-component of the velocity. Likewise, differentiating (3.28) and using (3.27) gives

$$\frac{d^2 v_y}{dt^2} = -\alpha^2\, v_y. \tag{3.32}$$

Both v_x and v_y are governed by the same differential equation. The general solutions are

$$\left. \begin{aligned} v_x &= A\,\cos(\alpha t) + C\,\sin(\alpha t) \\ v_y &= D\,\cos(\alpha t) + E\,\sin(\alpha t) \end{aligned} \right\}, \tag{3.33}$$

where A, C, D, and E are constants of integration (B is reserved for the magnetic field strength); we use different constants in the x and y directions as we eventually impose different boundary conditions on the two directions.

Integrating (3.33) with respect to time gives the equations of motion for the ion:

$$x = \frac{1}{\alpha}[A \sin(\alpha t) - C \cos(\alpha t)] + K_x \left.\right\}$$
$$y = \frac{1}{\alpha}[D \sin(\alpha t) - E \cos(\alpha t)] + K_y \left.\right\}, \qquad (3.34)$$

where K_x and K_y are further constants of integration.

Not all of A, C, D, and E are independent. This can be seen by back-substituting (3.33) into (3.27) [or into (3.28)—the result is the same]:

$$\frac{dv_x}{dt} = \alpha v_y$$
$$\Rightarrow -A \sin(\alpha t) + C \cos(\alpha t) = D \cos(\alpha t) + E \sin(\alpha t). \qquad (3.35)$$

This shows that we must have $D = C$ and $E = -A$. These constraints simplify (3.33) and (3.34) to

$$v_x = A \cos(\alpha t) + C \sin(\alpha t) \left.\right\}$$
$$v_y = C \cos(\alpha t) - A \sin(\alpha t) \left.\right\} \qquad (3.36)$$

and

$$x = \frac{1}{\alpha}[A \sin(\alpha t) - C \cos(\alpha t)] + K_x \left.\right\}$$
$$y = \frac{1}{\alpha}[C \sin(\alpha t) + A \cos(\alpha t)] + K_y \left.\right\}. \qquad (3.37)$$

We now set some initial conditions and impose them on (3.36) and (3.37). Assume that at $t = 0$ the positively-charged ion enters the magnetic field at $r_{initial} = (0, 0)$ while moving straight upward in the positive-y direction with velocity $v_{initial} = (0, v)$. This initial velocity can be supplied by passing the ions through an accelerating voltage before they are introduced into the magnetic field. The initial situation is sketched in Fig. 3.6.

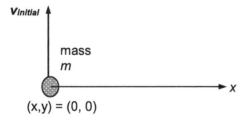

Fig. 3.6 A positively-charged ion is launched with initial velocity in the y direction; the magnetic field emerges from the plane of the page

The initial-velocity condition requires $A = 0$ and $C = v$ from (3.36); these results and the initial-position condition, when substituted into (3.37), demand $K_x = v/\alpha$ and $K_y = 0$. The velocity and position equations hence become:

$$\left.\begin{array}{l} v_x = v \, \sin(\alpha t) \\ v_y = v \, \cos(\alpha t) \end{array}\right\} \tag{3.38}$$

and

$$\left.\begin{array}{l} x = \dfrac{v}{\alpha}[1 - \cos(\alpha t)] \\[2mm] y = \dfrac{v}{\alpha} \, \sin(\alpha t) \end{array}\right\}. \tag{3.39}$$

Equations (3.38) indicate that an ion's speed remains unchanged once it enters the magnetic field; a magnetic field can do no work on a charged particle (why?). That (3.39) corresponds to circular motion can be appreciated by transforming to a new ("primed") coordinate system where the origin is displaced along the x-axis by an amount v/α:

$$\left.\begin{array}{l} x' = x - v/\alpha \\ y' = y \end{array}\right\}. \tag{3.40}$$

In this coordinate system, Eqs. (3.39) transform to

$$\left.\begin{array}{l} x' = -\dfrac{v}{\alpha} \, \cos(\alpha t) \\[2mm] y' = +\dfrac{v}{\alpha} \, \sin(\alpha t) \end{array}\right\}. \tag{3.41}$$

These expressions correspond to clockwise circular motion of radius v/α. The resulting motion is illustrated in Fig. 3.7.

From the definition of α, the radius of the orbit will be

$$R = \frac{v}{\alpha} = \frac{mv}{qB}. \tag{3.42}$$

The initial velocity v is usually created by accelerating the ions through an accelerating voltage V_{acc} before injecting them into the magnetic field. The resulting speed is given by

$$\frac{1}{2}mv^2 = qV_{acc} \Rightarrow v = \sqrt{\frac{2qV_{acc}}{m}}. \tag{3.43}$$

The orbital *diameter* $2R$ is then

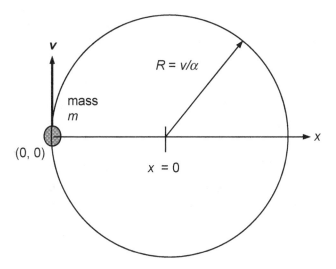

Fig. 3.7 Motion of a positively-charged charged particle in a magnetic field which emerges perpendicularly from the page. *v* is the velocity of the particle at the moment shown

$$D = \sqrt{\frac{8V_{acc}}{qB^2}} \sqrt{m}. \tag{3.44}$$

Heavier ions will have larger orbital radii; two ions of different mass entering the magnetic field will follow paths as sketched in Fig. 3.8.

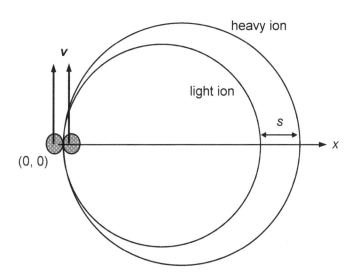

Fig. 3.8 As Fig. 3.7 but for ions of different masses

The ions will be maximally separated when they return to the x-axis after one-half of an orbit. The separation will be the difference of the diameters:

$$s = F\left(\sqrt{m_{heavy}} - \sqrt{m_{light}}\right), \quad F = \sqrt{\frac{8V_{acc}}{qB^2}}. \tag{3.45}$$

Hewlett and Anderson (1962, pp. 142–145) state that the Y-12 magnets at Oak Ridge produced a field of 0.34 T and that uranium tetrachloride (UCl$_4$) ion beams were accelerated to 35,000 Volts before being injected into the field. If the UCl molecules were singly ionized, (3.45) gives $F \sim 3.89 \times 10^{12}$ m kg$^{-1/2}$. The molecular weights of ^{235}UCl$_4$ and ^{238}UCl$_4$ are 375 and 378 mass units, respectively. Either of these values, when substituted into (3.44) gives a beam diameter of 3.07 m, and (3.45) gives a separation between the light and heavy-ion beams of 1.23 cm, or about half an inch.

For various reasons, the ion beam current represented by the streams of ^{235}UCl$_4$ ions in the Y-12 magnets had to be held to only a few hundred *micro*amperes (Parkins 2005). A beam current of 500 μA would correspond to collecting some 3.12×10^{15} ions per second. With a per-atom mass of 3.90×10^{-25} kg for ^{235}U, this means that one could collect some 1.22×10^{-9} kg of ^{235}U per second, or about 105 mg per day. To collect 50 kg at this rate would require some 1300 years of operation. It is thus understandable why the Y-12 facility eventually involved 1152 vacuum tanks, each utilizing two or four ion sources.

Some of the Y-12 magnets were square coils of about 30 windings and side lengths of 3 m (Reed 2009). The field at the center of such a coil is

$$B = \frac{2\sqrt{2}\,\mu_o N i}{\pi L}, \tag{3.46}$$

where $\mu_o = 4\pi \times 10^{-7}$ (Tesla-meter)/amp, L is the side length, i is the current, and N is the number of windings. With $L = 3$ meters and $N = 30$, the current required to generate a field of 0.34 Teslas is

$$i = \frac{\pi\,B\,L}{2\sqrt{2}\,\mu_o\,N} = \frac{\pi\,(0.34\,\text{T})\,(3\,\text{m})}{2\sqrt{2}\left(4\pi \times 10^{-7}\,\text{T} - \text{m/amp}\right)(30)} \sim 30,000\,\text{amp}. \tag{3.47}$$

In actuality, the current requirement was not this great; a history of the calutron program records that the magnets operated at between 4,000 and 7500 amperes—impressive figures nevertheless (Compere and Griffith 1991). Vacuum tanks through which the ion steams traveled were sandwiched between magnet coils; a given tank would have experienced fields from a number of neighboring coils. The Y-12 electromagnets were enormously consumptive of electricity, however. By July, 1945, the Y-12 facility had consumed some 1.6 billion kWh of electricity to enrich uranium for the *Little Boy* bomb. This amount of energy corresponds to about 1400 kilotons of TNT—some 100 times the yield of *Little Boy* itself! In the summer of 1945, Oak Ridge was consuming close to 1% of all the electricity being generated in the United States.

3.5 Gaseous (Barrier) Diffusion

Like electromagnetic separation, gaseous diffusion played a central role in enriching uranium for the *Little Boy* fission bomb. The physical principle utilized in this facility, which was code-named K-25, was that when a gas of mixed isotopic composition is pumped against a barrier made of a mesh of millions of tiny holes, atoms of the lighter isotope will tend to *diffuse* through the barrier slightly more readily than those of the heavier one (Strictly, the correct name for this process is *effusion*). The gas on the other side of the barrier, slightly enriched in the lighter isotope, is collected with a vacuum pump. However, the enrichment realizable through any one stage of barrier is limited by the relative masses of the two isotopes; the process must be repeated hundreds or thousands of times to achieve significant overall enrichment. In the case of uranium this is particularly so as the isotopes differ in mass by only about 1.3%. In fact, the input material to the K-25 plant was uranium hexafluoride gas, for which the isotopes differ by <1% in mass: $^{235}UF_6$ has atomic weight 349, while that of $^{238}UF_6$ is 352.

In view of the importance of gaseous diffusion to the success of the Manhattan Project, the physics of this process is derived here from first principles.

We begin with a result from classical thermodynamics. Suppose that we are dealing with a gas of atoms, each of mass m trapped in a container at absolute temperature T. According to the Maxwell-Boltzmann distribution, the mean atomic speed is given by

$$\langle v \rangle = \sqrt{\frac{8 k_B T}{\pi m}}. \tag{3.48}$$

We can imagine all atoms to have this speed, racing about in all possible directions. As shown in Fig. 3.9, imagine an abstract three-dimensional space where the axes are the (x, y, z) components of an atom's velocity. The magnitude of the velocity vector v shown in the diagram is v and its direction is given by spherical coordinates (θ, ϕ).

If there is no preferred direction of motion, then any direction of travel (θ, ϕ) must be as probable as any other. The solid angle subtended by angular limits θ to $\theta + d\theta$ and ϕ to $\phi + d\phi$ is $d\Omega = \sin\theta \, d\theta \, d\phi$; if (θ, ϕ) are measured in radians then the solid angle is said to be measured in steradians. Integrating overall all possible directions $[\theta = (0, \pi); \phi = (0, 2\pi)]$ shows that the total available solid angle is 4π steradians.

The probability that any atom chosen at random is moving in the direction of a particular solid angle $d\Omega$ is then given by $P(d\Omega) = d\Omega/4\pi$, that is,

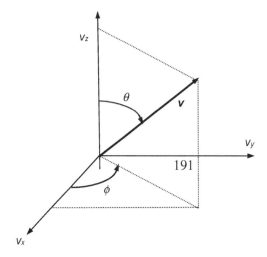

Fig. 3.9 Spherical coordinates. The axes are the components of the velocity vector v

$$P\,(d\Omega) = \frac{1}{4\pi}\,\sin\theta\,\,d\theta\,\,d\phi. \tag{3.49}$$

We now consider the diffusion process itself. Figure 3.10 shows a small portion of a diffusion barrier with a single hole of area S. In reality, there would be millions of such holes, but analyzing one of them will get us what we need. Atoms are pumped against the lower side of the barrier. All atoms are presumed to be moving at speed $\langle v \rangle$, and, at the moment shown, atom number 3 is just escaping through the hole. The fundamental problem is to compute the number of atoms that escape through the hole over some elapsed time Δt.

Over time Δt, an atom moving at speed $\langle v \rangle$ would travel distance $\langle v \rangle \Delta t$. As can be imagined with the aid of Figs. 3.10 and 3.11, any atom moving in the same direction as #3 and that is within an "escape cylinder" of slant length $\langle v \rangle \Delta t$ that projects back from the hole along the direction of v must escape within time Δt.

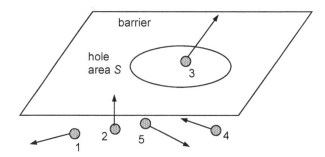

Fig. 3.10 Atoms moving in the vicinity of a hole of area S. Atom #3 is just escaping through the hole

Fig. 3.11 Escape cylinder
for a particle with velocity v

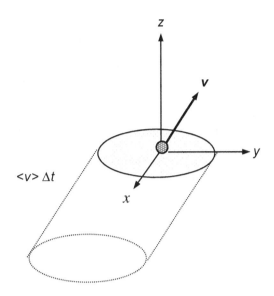

The number of atoms contained within the cylinder shown in Fig. 3.11 will be the volume of the cylinder times the number density of atoms $\rho_N = N/V$, where N is the number of atoms in the gas and V is the volume of the container (In this section, ρ designates number density, not mass density). To make number density a meaningful concept, we have to assume that the density of the gas stays constant as atoms fly in and out through the sides of the cylinder; we presume that for each atom that leaves the escape cylinder, one arrives to take its place.

The volume of a cylinder of top area S, slant length $\langle v \rangle \Delta t$, and tilt angle θ is given by

$$V_{cyl} = S \langle v \rangle (\Delta t) \cos \theta. \tag{3.50}$$

The number of atoms in the escape cylinder will then be

$$N_{cyl} = \rho_N S \langle v \rangle (\Delta t) \cos \theta. \tag{3.51}$$

Now, not all of these N_{cyl} atoms will be moving in the correct direction (θ, ϕ) to achieve escape. To account for this, we have to multiply (3.51) by the probability of an atom having its velocity so directed, which is given by (3.49):

$$N_{esc}(\Delta t, \theta, \phi) = N_{cyl} P (d\Omega) = \frac{\rho_N S \langle v \rangle (\Delta t)}{4\pi} \cos \theta \, \sin \theta \, d\theta \, d\phi. \tag{3.52}$$

We can account for all possible directions of escape by integrating (3.52) over the relevant angles:

$$N_{esc}(\Delta t) = \frac{\rho_N S \langle v \rangle (\Delta t)}{4\pi} \int\limits_0^{\pi/2} \sin\theta \, \cos\theta \, d\theta \int\limits_0^{2\pi} d\phi. \tag{3.53}$$

Notice that that the limits on θ here run from 0 to only $\pi/2$ (not π); we want to account for only *outward*-moving atoms. Since the diffusion barrier is packed with millions of holes practically edge-to-edge, it will not matter if an atom is offset from the one shown in the figures; any outward-moving atom will find a hole to escape through.

The integrals appearing in (3.53) evaluate to $1/2$ and 2π. Combining these with (3.48) gives the important result

$$N_{esc}(\Delta t) = \frac{1}{4}\rho_N S\langle v \rangle(\Delta t) = C\left(\frac{\rho_N}{\sqrt{m}}\right), \tag{3.54}$$

where

$$C = (S\,\Delta t)\sqrt{\frac{k_B T}{2\pi}}. \tag{3.55}$$

Equation (3.54) is the central result for understanding barrier diffusion; it tells us that the number of atoms destined to escape through a hole of area S over time Δt is proportional to their number density, and inversely proportional to the square root of their mass; S could in fact as well represent the area of all of the holes in the barrier. This equation also plays a central role in the derivation of the neutron diffusion equation in Appendix G.

Now consider a gas consisting of a single-isotope species. All stages of the diffusion mechanism are presumed to have the same volume V, the same hole area S, and to operate at the same temperature T for the same time Δt; that is, that the constant C is presumed to be the same for each stage of the diffusion cascade. Let ρ_o be the number density of the feedstock to the first stage of the cascade. From (3.54), the number of atoms that escape from the first stage of the diffuser will be

$$N_1 = C\left(\frac{\rho_o}{\sqrt{m}}\right). \tag{3.56}$$

The number density of atoms in the second stage will then be N_1/V, or

$$\rho_{\substack{enter \\ stage\,2}} = \frac{N_1}{V} = \frac{C}{V}\left(\frac{\rho_o}{\sqrt{m}}\right). \tag{3.57}$$

With this input number density for stage 2, the number of atoms that escape through stage 2 is given by re-applying (3.54):

$$N_2 = C \left(\frac{\rho_{enter \atop stage\ 2}}{\sqrt{m}} \right) = \frac{C^2}{V} \frac{\rho_o}{\left(\sqrt{m}\right)^2}. \tag{3.58}$$

Propagating this logic shows that after a total of n successive stages, the number of atoms that emerge from the n'th stage will be

$$N_n = \frac{C^n}{V^{n-1}} \frac{\rho_o}{\left(\sqrt{m}\right)^n}. \tag{3.59}$$

If the gas consists of a mixture of two isotopes, say ^{235}U and ^{238}U, (3.59) will apply to each according as the relevant values of ρ_o and m. If we designate the two isotopes with subscripts 5 and 8, then the final ratio of the number of 235 atoms to the number of 238 atoms can be written as

$$\frac{N_5}{N_8} = \left(\frac{\rho_{o5}}{\rho_{o8}} \right) \left(\frac{m_8}{m_5} \right)^{n/2}. \tag{3.60}$$

Even if different stages of the cascade have different values of V, S, T or Δt, (3.60) will still be correct as it is formulated as a ratio, and those quantities will cancel at each stage since they apply equally to each isotope.

Since $m_8 > m_5$, (3.60) indicates that the ratio N_5/N_8 grows with each stage. However, the amount of enrichment achieved at each stage is tiny: If we start with uranium of natural isotopic composition and ignore the small natural abundance of ^{234}U, $\rho_{o5}/\rho_{o8} = 0.0072/0.9928 = 7.25 \times 10^{-3}$, and, with uranium hexafluoride, $m_8/m_5 = 1.0086$.

The extent of enrichment is usually quantified by the percentage of ^{235}U. If we define $x = N_5/N_8 = \rho_5/\rho_8$, then

$$\% (235) = 100 \left(\frac{x}{x+1} \right). \tag{3.61}$$

Figure 3.12 shows the run of percent ^{235}U as a function of the number of diffusion stages, assuming that one starts with uranium of natural isotopic composition.

Bomb grade ^{235}U is usually considered to be reached at 90% enrichment ($x = 9$), which requires $n = 1665$. In the case of 1000 stages, 34% enrichment can be realized, whereas 50% enrichment requires $n = 1151$. The K-25 plant comprised 2892 stages, which would theoretically have realized 99.94% enrichment, but in actuality the feed material was input about one-third of the way along the cascade so that "depleted" uranium hexafluoride could be recycled to preceding stages. Figure 3.13 shows a schematic illustration of a section of a diffusion "cascade".

At Oak Ridge, uranium went through various stages of enrichment in various facilities as they were brought into service. When all enrichment methods had come on-line by the spring of 1945, natural-abundance uranium hexafluoride was first fed into the liquid thermal diffusion plant (code-named S-50), which enriched the

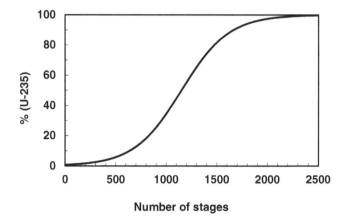

Number of stages

Fig. 3.12 ^{235}U enrichment as a function of number of diffusion stages. The initial isotopic abundance is assumed to be that of natural uranium

Fig. 3.13 Schematic illustration of a diffusion cascade. Feed material enters the cascade in the second "cell" from the bottom of the diagram. The dashed lines inside each cell represent the diffusion membrane, and the circles represent pumps. Gas enriched in the lighter isotope accumulates toward the top of the diagram, while that depleted in the lighter isotope accumulates toward the bottom. In reality, the cascade is not arranged vertically as this diagram suggests; in the K-25 plant, all cells were at ground level. Sketch by author

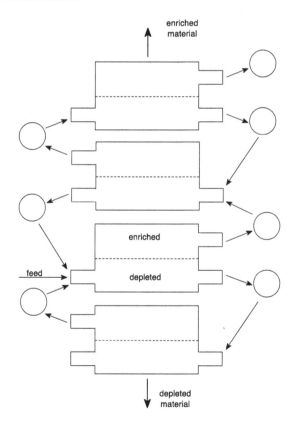

^{235}U content from 0.72 to 0.86%. This product would be fed to first-stage calutrons, known as "Alpha" calutrons, which could raise the enrichment level to about 20%. Second-stage "Beta" calutrons, which were smaller, took the enrichment to 90%. When enough K-25 stages were on-line to produce 20% enriched material directly, the original Alpha units were shut down, and K-25's product was directed to "second-generation" Alpha units and thereafter to Beta units. K-25's enrichment level peaked at 36%, although postwar extensions to the plant brought the level up to bomb-grade enrichment. Enriched uranium tetrachloride from the calutrons was converted to uranium hexafluoride for shipment to Los Alamos, with chemical processing carried out in gold trays to minimize contamination.

By the end of the war, the continuous-feed K-25 gaseous diffusion process had proven itself more efficient at enriching uranium than the electromagnetic method; shutdown of Alpha calutrons began on September 4, 1945. Some Beta calutrons remained in operation until 1998 to separate various isotopes, which, after neutron bombardment in a reactor, could be used as radioactive tracers for medical-imaging and cancer-treatment applications. The K-25 plant (plus other similar plants) continued to operate to produce both highly-enriched and low-enriched uranium at Oak Ridge until it was shut down in 1985; by this time, centrifugation had become a more economical means of enrichment. The last remnants of K-25 were demolished in early 2013.

According to a 2011 report by the International Panel on Fissile Materials, the United States produced a total of some 610 metric t (610,000 kg) of highly-enriched uranium between 1945 and 1995 (Global Fissile Material Report 2011).

References

Compere, A.L., Griffith, W.L.: The U.S. calutron program for uranium enrichment: history, technology, operations, and production. Oak Ridge National Laboratory report ORNL-5928 (1991)

Fermi, E.: Experimental production of a divergent chain reaction. Am. J. Phys. **20**, 536–558 (1952)

Garwin, R.L., Charpak, G.: Megawatts and megatons: a turning point in the nuclear age? Alfred A. Knopf, New York (2001)

Global Fissile Material Report. International Panel on Fissile Materials (2011). http://fissilemater ials.org/library/gfmr11.pdf

Hewlett, R.G., Anderson, O.E.: The New World 1939/1946: Volume I Of A History of the United States Atomic Energy Commission. The Pennsylvania State University Press, University Park, PA (1962)

Historic American Engineering Record: B Reactor (105-B) Building, HAER No. WA-164. DOE/RL-2001-16. United States Department of Energy, Richland, WA (2001). http://wcpeace. org/history/Hanford/HAER_WA-164_B-Reactor.pdf

Jones, R.C., Furry, W.H.: The separation of isotopes by thermal diffusion. Rev. Mod. Phys. **18**, 151–224 (1946)

Parkins, W.E.: The uranium bomb, the calutron, and the space-charge problem. Phys. Today **58**(5), 45–51 (2005)

Reed, B.C.: Understanding plutonium production in nuclear reactors. Phys. Teach. **43**, 222–224 (2005)

Reed, B.C.: Bullion to B-fields: the silver program of the Manhattan Project. Michigan Academician **XXXIX**(3), 205–212 (2009)

Reed, B.C.: Liquid thermal diffusion during the Manhattan Project. Phys. Perspect. **13**(2), 161–188 (2011)

Reed, B.C.: Rousing the dragon: polonium production for neutron generators in the Manhattan Project. Am. J. Phys. **87**(5), 377–383 (2019)

Weinberg, A.M.: Eugene Wigner, nuclear engineer. Phys. Today **55**(10), 42–46 (2002)

Chapter 4
Complicating Factors

A number of complications can thwart the proper functioning of nuclear weapons. In this chapter we explore several of these. In the Manhattan Project, graphite was used as a moderating medium in plutonium-production reactors, and in Sect. 4.1 we analyze the difficulties of operating such a reactor in the face of effects of the neutron-capturing properties of graphite itself and any impurities that might be present. The other two factors involve the problem that if stray neutrons should be present during the brief time that one is assembling sub-critical pieces of fissile material to form a supercritical core, one runs the risk that such neutrons could initiate a premature detonation. Some of these "background" neutrons arise from spontaneous fissions within the fissile material itself and are fundamentally uncontrollable, while others arise from so-called (α, n) reactions due to the presence of light-element impurities in the fissile material. The issue of spontaneous fission is taken up in Sects. 4.2 and 4.3, where the probability of a spontaneous fission-initiated "fizzle" and the expected yield of a weapon in such a case are examined. The light-element issue, where the creation of neutrons can be minimized (albeit with difficulty) is examined in Sect. 4.4. Also, it is essential to provide a way of triggering a nuclear explosion once a suitable assembly of fissile material has been achieved. In the Manhattan Project, this was achieved with so-called "neutron initiators"; ironically, these devices functioned via the same (α, n) reactions which had otherwise to be avoided. The physics and production of initiator is explored in Sect. 4.5. Finally, some of the yield of the *Trinity* bomb was due to high-energy neutrons inducing fission in the ^{238}U tamper shell of that device. This effect is not in any sense a malfunction, but does complicate the question of estimating a bomb's yield. This is analyzed in Sect. 4.6.

© The Author(s), under exclusive license to Springer Nature Switzerland AG 2021   145
B. C. Reed, *The Physics of the Manhattan Project*,
https://doi.org/10.1007/978-3-030-61373-0_4

4.1   Boron Contamination in Graphite

The presence of impurities in the graphite used as a moderator in the CP-1 and Hanford reactors was a matter of serious concern in the Manhattan Project. Since the purpose of the graphite was to slow and scatter neutrons without capturing them, it was important for it to be as free as possible of any neutron-capturing impurities. Some neutron capture was inevitable as carbon itself has a small capture cross-section for thermal neutrons (3.53 millibarns), but the greater danger was that commercially-produced graphite at the time often contained trace amounts of boron, which has a voracious appetite for capturing neutrons. Indeed, it was unappreciated boron contamination of graphite that led German researchers to conclude that only heavy water could serve as an adequate moderator, a decision that was at least in part responsible for their failure to achieve a self-sustaining chain-reaction during World War II (Reed 2020). In this section we examine the severity of this effect.

Two isotopes of boron occur naturally: ^{10}B (19.9%) and ^{11}B (80.1%). Boron-10 has a minute thermal-neutron capture cross-section, but that of boron-10 is enormous, about 3840 barns. Weighting by abundance, this gives boron a bulk capture cross-section of about 760 barns: A single "average" atom of boron has a neutron capture effect equivalent to that of over 200,000 carbon atoms. The culprit reaction involved is neutron capture to produce an alpha particle, an (n, α) process:

$$\, _0^1 n \, + \, _5^{10}B \, \rightarrow \, _2^4He \, + \, _3^7Li. \tag{4.1}$$

As will be seen, the presence of even a small amount of boron-10 can quickly suppress the desired chain reaction.

The approach used in Sect. 3.1 to investigate reactor criticality can be modified to account for capture effects due to carbon and boron. As in that section, let σ_{f5}, σ_{c5}, and σ_{c8} designate the cross-sections for fission and capture by ^{235}U and capture by ^{238}U. To these, add symbols for capture by carbon and boron: σ_{cC} and σ_{cB}.

Imagine the pile idealized as a homogeneous mixture of uranium, carbon, and boron atoms, with thermalized neutrons flying about. The minimal graphite-to-carbon ratio necessary to sustain a chain reaction is a matter of nuclear engineering, but I make an estimate here by assuming knowledge of a graphite pile that is known to have just achieved criticality: Enrico Fermi's CP-1. From figures given in a paper published by Fermi on the tenth anniversary of that achievement [Fermi (1952)], it can be determined that CP-1 contained ~37,700 kg of pure uranium and ~349,700 kg of graphite, which gives a C:U mass ratio of ~9.3:1. The molecular weight of uranium is about 19.8 times that of carbon, so this corresponds to a C:U number ratio of ~180:1. Call this ratio R. Analogous numbers can be computed for the Oak Ridge X-10 and Hanford reactors, and indicate $R \sim 120$, but those devices are complicated by the presence of fuel channels and coolant. I will stick with $R = 180$; the results which follow are not, however, wildly sensitive to reasonable changes in R.

Let the number of boron atoms per carbon atom be B. Then the number of boron atoms per uranium atom will be BR, and, in analogy to (3.3), we can write the total effective cross section per atom as

$$\sigma_{total} = F(\sigma_{c5} + \sigma_{f5}) + (1 - F)\sigma_{c8} + R\sigma_{cC} + BR\sigma_{cB}, \qquad (4.2)$$

where F is again the fractional abundance of ^{235}U. As before, the total fission cross section per atom will be $F\sigma_{f5}$, so the probability that a neutron which strikes any nucleus will induce a fission will be $F\sigma_{f5}/\sigma_{total}$. If each fission liberates ν neutrons, then the reproduction factor k becomes, in analogy to (3.4),

$$k = \frac{\nu F \sigma_{f5}}{\sigma_{total}} = \frac{\nu F \sigma_{f5}}{F(\sigma_{c5} + \sigma_{f5}) + (1 - F)\sigma_{c8} + R\sigma_{cC} + BR\sigma_{cB}}. \qquad (4.3)$$

With the values given above and in Table 3.1, we can used this to estimate the maximum tolerable number of boron atoms per carbon atom to keep $k > 1$. With $F = 0.0072, R = 180, \sigma_{cC} = 0.00353$, and $\nu = 2.4$, we find $B = 1.35 \times 10^{-5}$, which means only one boron atom per 74,000 carbon atoms. Given the a real reactor is bound to suffer other surces of neutron loss (fuel tubes and cladding, coolant, containment structures, control rods), this would be the best-case scenario. If these effcets reduce ν to an effective value of 2, only one boron per 837,000 carbons can be tolerated. Other possible contamimants can come into play as well. For example, nitrogen has a neutron capture cross section of about 75 millibarns, some 20 times that of carbon; one must be careful not let air become trapped between layers of graphite. The moral of the story is that reactor-grade graphite must be kept very pure; according to a U. S. Department of Energy history, boron in the graphite blocks of the Hanford reactors was held to a level of 0.4 parts per million (DOE 2001).

It was remarked at the beginning of this section that unappreciated boron contamination of graphite led German researchers, led by then-future Nobel Laureate Walther Bothe (of the neutron discovery history related in Sect. 1.4) to reject that material as a moderating medium. Bothe and his collaborators, assuming that their graphite was pure, measured the neutron capture cross-section of carbon, which they determined to be ~6.4 millibarns as opposed to the correct value of 3.53 millibarns. In this case, it will be impossible to achieve $k = 1$ for $\nu = 2$ even if no boron is present. Ironically, a team led by Enrico Fermi was undertaking similar measurements in the United States, and with their purer graphite determined a cross section of about 3 millibarns. At the time, neither group openly published their results. Bothe would surely have revisited his work had Fermi done so, with consequences which might have been dire.

4.2 Spontaneous Fission of ^{240}Pu, Predetonation, and Implosion

Material in this section is adopted from Reed (2010).

Emilio Segrè's discovery in December, 1943, that ^{235}U has a very low spontaneous fission (SF) rate cleared the way for that material's use in the "gun assembly" mechanism of the *Little Boy* bomb. Conversely, his later discovery that reactor-produced plutonium has a very high SF rate meant that a gun assembly method would be far too slow for the *Trinity* and *Fat Man* bombs. The problem was not with the ^{239}Pu to be used as fissile material for the bombs, but rather that some ^{240}Pu was inevitably formed in the Hanford reactors as a consequence of already-formed ^{239}Pu nuclei capturing neutrons. ^{240}Pu has an extremely high SF rate, and only implosion could trigger a plutonium bomb quickly enough to prevent a SF from causing a premature detonation. In this section we examine the probability of predetonation; in Sect. 4.3 we look at a model for estimating what fraction of a bomb's design yield we might expect to realize given the possibility of predetonation. How one can estimate the amount of ^{240}Pu created in a reactor is analyzed in Sect. 5.3.

Let A designate the atomic weight (gr mol^{-1}) of some spontaneously fissioning material. The number of atoms in one kilogram of material will then be $10^3(N_A/A)$. For any decay process characterized by a half-life $t_{1/2}$ s, the average lifetime is $t_{1/2}/(\ln 2)$. Consequently, the average spontaneous fission rate F (number per kilogram per second) is given by the number of nuclei divided by their average lifetime:

$$F = 10^3 \left(\frac{N_A}{A} \right) \left(\frac{\ln 2}{t_{1/2}} \right) (\text{kg}^{-1} \, \text{s}^{-1}). \tag{4.4}$$

Recommended values for SF half-lives for heavy isotopes have been published by Holden and Hoffman (2000). Numbers for four isotopes of interest are given in Table 4.1. The spontaneous fission rates in the fourth column of the Table are quoted in number per kilogram of material per 100 μs. The secondary-neutron v values for ^{238}U and ^{240}Pu represent the number of neutrons emitted in spontaneous fissions of these nuclides; these are adopted from Table 1.33 of Hyde (1964).

The reason for quoting the SF rates per 100 μs goes back to the design of the *Little Boy* bomb. For a core on the order of 10 cm in size assembled at 1000 m/s, about 100 μs will be required to complete the assembly. During this time, a 50-kg ^{235}U

Table 4.1 Spontaneous fission parameters

Nuclide	$t_{1/2}$ (year)	A (gr/mol)	SF (kg 100 μs)$^{-1}$	v
^{235}U	1.0×10^{19}	235.04	5.627×10^{-7}	2.637
^{238}U	8.2×10^{15}	238.05	6.776×10^{-4}	2.1
^{239}Pu	8×10^{15}	239.05	6.916×10^{-4}	3.172
^{240}Pu	1.14×10^{11}	240.05	48.33	2.257

assembly would suffer some 2.81×10^{-5} spontaneous fissions; the probability of predetonation would be miniscule (although not zero). Also, contamination of a few percent ^{238}U in a ^{235}U core will not present a significant hazard as far as spontaneous fissions are concerned. Similarly, for a *pure* 10-kg ^{239}Pu core, the rate is about 0.007 spontaneous fissions per 100 μs. However, a 10-kg plutonium core contaminated with even only 1% ^{240}Pu is likely to suffer some 5 spontaneous fissions during this brief time; the core pieces are unlikely to reach their fully assembled configuration before a spontaneous fission causes a pre-detonation. The only option, aside from the virtually impossible task of trying to remove the offending ^{240}Pu, is to speed up the assembly process to on the order of a microsecond.

While the above numbers give a sense of the potential magnitude of the possibility of a SF-induced predetonation, a more careful analysis is necessary to fully quantify this risk. Because spontaneous fission is a random phenomenon, one is restricted to speaking in terms of probabilities. The physics of the situation will dictate a certain probability that a predetonation may happen; it is then a judgment call as to the acceptability of that risk.

The approach taken here is based on a probabilistic model of neutrons traveling through a bomb core, and should be understandable to readers familiar with the concept of multiplying together independent probabilities to generate an overall probability. To calculate the predetonation probability, we have to treat two effects: (i) The probabilities that 0, 1, 2, ... spontaneous fissions occur during the assembly time; and (ii) The probability that the secondary neutrons so released travel to the edge of the core and escape without causing secondary fissions.

Imagine a spherical bomb core containing mass M of spontaneously fissioning material, and let F be the rate of spontaneous fissions as given by (4.4). We assume a spherical geometry for the bomb core while it is being assembled—an obviously somewhat unrealistic model for a gun-type bomb. The average number of spontaneous fissions during the assembly time $t_{assemble}$ will be

$$\mu = M F t_{assemble}. \tag{4.5}$$

From Poisson statistics, the probability P_k ($k = 0, 1, 2, ...$) that exactly k spontaneous fissions occur during this time is given by

$$P_k = \frac{\mu^k}{k!} e^{-\mu}. \tag{4.6}$$

If each spontaneous fission releases on average ν neutrons, then k spontaneous fissions will release $k\nu$ neutrons. For no predetonation to occur, all of these neutrons must escape. If P_{escape} represents the probability that an individual neutron will escape without causing a fission, then the probability that all will escape is $\left(P_{escape}\right)^{k\nu}$. Hence, the probability that both k spontaneous fissions occur and that all of the emitted neutrons escape is $P_k \left(P_{escape}\right)^{k\nu}$. How P_{escape} is determined is described in the following paragraphs.

To determine the probability of no predetonation, we have to account for all possible number of occurrences of spontaneous fissions:

$$P_{\substack{no \\ predet}} = \sum_{k=0} P_k \left(P_{escape}\right)^{kn}. \tag{4.7}$$

In principle, the sum here goes to infinity, but in practice the first few terms suffice because P_k in (4.6) declines very quickly with increasing k due to the factorial term.

The next part of the argument is to determine P_{escape}, the overall escape probability for a single neutron.

Neutrons can escape the core in one of two ways: they may escape directly by traveling in a straight line from their point of origin to the edge of the sphere, or they may scatter one or more times before escaping. For a given neutron, it is impossible to predict how many times it will scatter before escaping, but we can develop an expression for the probability that it will escape following a specified number of scatterings; adding these probabilities gives P_{escape}. It is useful to imagine that S_{max}, the maximum possible number of scatterings before escape, is known in advance. How S_{max} is estimated is discussed following (4.13) below.

If π_j represents the probability that a neutron escapes following j successive scatterings, then the overall total probability of escape is

$$P_{escape} = \pi_0 + \pi_1 + \pi_2 + \cdots + \pi_{S_{max}}. \tag{4.8}$$

To determine the π_j, recall the expression from Sect. 2.1 for the probability that a neutron will penetrate through a linear distance x of material: $P(x) = \exp(-\sigma_{total}nx)$, where n is the number density of nuclei in the material and σ_{total} is the total reaction cross-section for neutrons against the material. As in the calculation of critical mass, σ_{total} is given by the sum of the scattering and fission cross sections, since *any* type of interaction must be avoided if a neutron is to escape directly. We ignore any possibility of non-fission neutron capture, which for any reasonably pure fissile material should be small. Now, this $P(x)$ refers to neutrons penetrating through a linear distance x. If the neutrons are emitted in random directions within the bomb core, we need to average $P(x)$ over all possible directions of neutron emission from all points within the sphere. So as not to disturb the flow of the present argument, this issue is examined in Appendix F, where it is shown that the appropriate average, $\langle P_{sph} \rangle$, can be expressed a very compact analytic form

$$\langle P_{sph} \rangle = \frac{3}{8x^3}\left[2x^2 + e^{-2x}(2x+1) - 1\right], \tag{4.9}$$

where $x = \sigma_{total}\, nR_{core}$.

The probability that a neutron will not directly escape is $1 - \langle P_{sph} \rangle$. These neutrons must first interact with a nucleus by either causing a fission (f) or by being scattered (s). The respective probabilities of these competing processes are σ_f/σ_{total} and σ_s/σ_{total}. Hence, the probability that a neutron will suffer one scattering is given by

$$P_{one} = \left(\frac{\sigma_s}{\sigma_{total}}\right)\left(1 - \langle P_{sph}\rangle\right) \equiv g. \tag{4.10}$$

The probability that such a once-scattered neutron will then escape, that is, π_1 of (4.8), is given by P_{one} times $\langle P_{sph}\rangle$:

$$\pi_1 = \left(\frac{\sigma_s}{\sigma_{total}}\right)\left(1 - \langle P_{sph}\rangle\right)\langle P_{sph}\rangle = g\,\langle P_{sph}\rangle. \tag{4.11}$$

Similarly, the probability that a neutron that has already undergone one scattering will experience a second scattering is given by P_{one} of (4.10) times the probability of suffering a further interaction, $\left(1 - \langle P_{sph}\rangle\right)$, times the probability of that interaction being a scattering, σ_s/σ_{total}, that is, $P_{two} = g^2$. The probability of escape after two scatterings is thus $\pi_2 = g^2\,\langle P_{sph}\rangle$. Carrying on this logic and assuming that scatterings are independent events, the probability that a neutron will suffer j successive scatterings and then escape is given by

$$\pi_j = g^j\langle P_{sph}\rangle. \tag{4.12}$$

Hence we have

$$P_{escape} = \langle P_{sph}\rangle\left(\sum_{j=0}^{S_{max}} g^j\right) = \langle P_{sph}\rangle\left[\frac{1 - g^{S_{max}+1}}{1 - g}\right]. \tag{4.13}$$

The right side of Eq. (4.13) follows from the fact that the summation is the partial sum of a geometric series. We have assumed that neutrons are randomly redirected at each scattering, which is not strictly true.

What about the maximum number of scatterings S_{max}? The scenario which maximizes the predetonation probability, that is, the worst case, is $S_{max} = 0$. That the worst-case scenario corresponds to $S_{max} = 0$ may seem counterintuitive, as one might expect more neutron-nucleus interactions to lead to more chances for fissions. This is true, but some neutrons may escape even after a very large number of scatterings; setting $S_{max} = 0$ means that we forego accounting for such escapees, leading to an overestimate of the predetonation probability. In the spreadsheet developed to perform these calculations, the value of S_{max} is to be assigned by the user as desired. In any event, the precise choice of S_{max} proves not to be drastically significant; it is shown below that for a model of the *Little Boy* ^{235}U core, the predetonation probability changes by less than 1% for reasonable choices of S_{max}. For the *Fat Man* ^{239}Pu core the sensitivity is somewhat greater, but the value of S_{max} is by no means a determining factor in whether or not implosion is necessary.

Spreadsheets **PreDetonation(LB).xls** and **PreDetonation(FM).xls** have been developed to carry out these calculations for *Little Boy* and *Fat Man* configurations. The user enters the core and contaminant masses and their atomic weights, the relevant cross-sections, the SF half-life and secondary neutron number for the spontaneously fissile material, the maximum number of scatterings to be considered, and the assembly timescale. To calculate the sum in (4.13), the spreadsheets take an upper limit of $k = 20$, which is entirely sufficient for any reasonable situation. Results of such calculations are described in the following two subsections.

4.2.1 Little Boy Predetonation Probability

As described in the Preamble, the Hiroshima *Little Boy* core comprised about 66 kg of uranium in a cylindrical configuration, of which about 80% was ^{235}U and 20% (13.2 kg) was ^{238}U. The half-life of ^{238}U for spontaneous fission, 8.2×10^{15} year, is about 1200 times shorter than that of ^{235}U, rendering the latter isotope almost negligible as far as the predetonation probability is concerned. As in Sect. 2.3, I model the (bare) core of *Little Boy* as being spherical; a 66 kg sphere of density 18.71 gm cm^{-3} has a radius of 9.44 cm. For a 200 μs assembly time, an average of only 0.018 spontaneous fissions will occur; the probability that *no* spontaneous fissions will occur at all is 98.2%. The non-predetonation probability evaluates as about 98.3% for $S_{max} = 0$, and as 98.8% for $S_{max} = 5$ (probably too large). At worst, fizzles could be expected to occur in about two such bombs out of every one-hundred. The spherically-averaged direct escape probability $\langle P_{sph} \rangle$ for this 66 kg core is 0.265; for $S_{max} = 5$, P_{escape} of (4.13) is 0.606. For a 100 μs assembly time, the mean number of spontaneous fissions is only about 0.009, and the probability of no pre-detonation for $S_{max} = 0$ rises to 99.2% ($S_{max} = 0$).

4.2.2 Fat Man Predetonation Probability

The untamped critical mass of ^{239}Pu is about 17 kg. However, the *Trinity* and *Fat Man* bombs used cores of mass about 6.3 kg due to the greater efficiency afforded by implosion. For a 6.3 kg core of pure ^{239}Pu, an assembly time of 200 μs yields a no-predetonation probability of 99.2% ($S_{max} = 0$). In reality, however, the cores contained about 1.2% Pu-240 (0.0756 kg), which makes the outcome very different. Figure 4.1 shows the $S_{max} = 0$ non-predetonation probability for this case as a function of the assembly time. For $S_{max} = 0$ and a time of 100 μs, the non-predetonation probability is only 5.5% (11.5% for $S_{max} = 1$); there is consequently no realistic hope of successfully assembling the core in a time scale characteristic of a gun mechanism. Here $\langle P_{sph} \rangle = 0.496$, and, for $S_{max} = 5$, $P_{escape} = 0.770$. Although these numbers do not differ much from those of the *Little Boy* calculation, the mean number of spontaneous fissions is enormously greater in the case of the ^{240}Pu-contaminated *Fat Man* device: 3.7 in comparison to 0.009 over 100 μs.

We can make a rough estimate of the imploded *Trinity* core non-predetonation probability as follows. Neglecting the neutron initiator housed at its center, a 6.3-kg core would have a radius of about 4.59 cm for a density of 15.6 gm cm^{-3}. If the

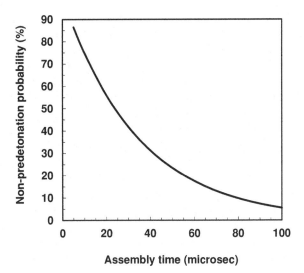

Fig. 4.1 Non-predetonation probability for a non-imploded 6.3-kg Pu core contaminated with 1.2% Pu-240 as a function of assembly time. The maximum number of scatterings is assumed to be zero

implosion crushes the core to a density twice this value, the final radius would be about 3.64 cm. If this is done at a speed of, say, 2000 m/s, some 4.7 μs would elapse. Modeling the core as having a density midway between these values, 23.4 gm cm^{-3}, gives an $S_{max} = 0$ non-predetonation probability of 86.3% (88.7% for $S_{max} = 1$) for 1.2% ^{240}Pu contamination for this elapsed time. The altered density actually makes little difference to the probabilities: at 15.6 gm cm^{-3} they are 87.2% and 90.4% for $S_{max} = 0$ and $S_{max} = 1$, respectively, for a time of 4.7 μs. These estimates accord very well with an analysis published in 2009 by former Los Alamos Theoretical Division Director Carson Mark: Working from figures given in a letter from Robert Oppenheimer to Manhattan Engineer District Commander General Leslie Groves, Mark reported that Oppenheimer estimated a 88% chance of a "nominal" 20 kiloton yield from the *Trinity* device (Mark et al. 2009).

The choice of 2000 m/s for an implosion speed is reasonable. The speed of a sound wave through a material is given by the formula $v = \sqrt{B/\rho}$, where B is the bulk modulus of the material and ρ its density. B has units of pressure, and is a measure of the compressibility of the material; it depends on factors such as the crystal structure of the material and the temperature. Various estimates of the B-value of plutonium can be found in the scientific literature, with $B \sim 50 \times 10^9$ Pa being a representative value. With $\rho \sim 15{,}600$ kg m^{-3} for uncompressed plutonium, $v \sim 1800$ m/s. This issue is explored further in Exercise 4.2.

Given that plutonium synthesized in fuel rods in commercial reactors comprises about 20% Pu 240, we can appreciate the difficulties faced by terrorists who would plan to steal spent fuel rods and use them to create a workable plutonium bomb. Mark concluded, however, that even a 0.5-kiloton "fizzle yield" for a terrorist bomb based on reactor-grade plutonium would still produce a severely damaging explosion. Such an explosion would be equivalent to about 200 of the truck bombs used to destroy the

Murrah Federal Building in Oklahoma City in 1995 (Bernstein 2008). Thus, while an efficient *Trinity*-like terrorist weapon based on purloined fuel rods is highly unlikely, the issue of fissile-material security will remain a pressing one for years to come.

Figure 4.2 shows a photograph of the *Trinity* test device; Fig. 4.3 shows the Nagasaki *Fat Man* bomb; the bulbous casing enclosed the implosion assembly within and provided stable flight characteristics after release from its bomber.

To close this section, we ask: "How can one obtain an *implosion*?" After all, explosions are normally seen to be outwardly-directed phenomena. In the Manhattan Project, this was achieved by using an assembly of *implosion lenses*. The fundamental idea is sketched in Fig. 4.4, which shows a single lens in cross-section; to extend the idea to three dimensions, imagine a somewhat pyramidal-shaped block that would fit comfortably on your lap. The block comprises two explosive castings that fit together very precisely. The outer casting is of a fast-burning explosive known as "Composition B" (or just Comp B), while the inner, lens-shaped one is a slower-burning material known as Baratol, a mixture of barium nitrate and TNT. A detonator at the outer edge of the Comp B initiates an outward-expanding detonation wave. When the detonation wave hits the Baratol, it too begins exploding. If the interface between the two is of just the right shape, the two waves can be arranged to combine as they progress along the interface in such a way as to create an inwardly-directed converging wave in the Baratol; the dashed lines in Fig. 4.4 illustrate the right-to-left progression of the detonation. As sketched in Fig. P.5 of the Preamble, 32 such "binary explosive" assemblies were fitted together to create an imploding sphere inside the *Trinity* and *Fat Man* devices. Within the Baratol lenses resided another spherical assembly of 32 blocks of Comp B, which are detonated by the Baratol to achieve a high-speed symmetric crushing of tamper spheres that lay within them. A very readable personal reminiscence of casting and machining implosion lenses was published by Hull and Bianco (2005); for a more technical history, see Hoddeson et al. (1993).

Fig. 4.2 The *Trinity* device atop its test tower on July 15, 1945. Norris Bradbury (1909–1997), who served as Director of the Los Alamos Laboratory from 1945 to 1970, stands to the right. The spherical shape of this implosion device is clearly visible; the cables feeding from the box halfway up the device go to the implosion-lens detonators discussed in the text. Photo courtesy Alan Carr, Los Alamos National Laboratory

Fig. 4.3 The Nagasaki *Fat Man* plutonium implosion weapon shortly before its mission. *Fat Man* was 12 feet long, 5 feet in maximum diameter, and weighed 10,300 lb when fully assembled (Sublette 2007). Photo courtesy Alan Carr, Los Alamos National Laboratory

Fig. 4.4 Schematic illustration of implosion lens segment. Sketch by author

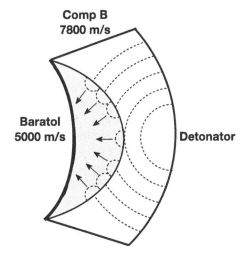

4.3 Predetonation Yield

Material in this section is adopted from Reed (2011).

In the previous section, it was explained how uncontrollable spontaneous fissions inevitably lead to some probability that a nuclear weapon will suffer a predetonation.

Given this situation, a corollary question arises: Is it possible to predict what fraction of the design yield of a weapon might be realized in the case of such an event? Historically, concern with the yield-fraction probability was motivated not only by the desire to have some idea of what yield might be expected, but also by the desire to ensure an explosion violent enough to destroy the bomb and disperse the fissile material even if a minimum-yield explosion occurred. The rationale for this is that if a bomb fails to operate properly but still destroys itself, an adversary would be unable to recover the fissile material and reverse-engineer the weapon. A minimum-yield explosion is known to weapons engineers as a "fizzle."

A yield-fraction model comprises two separate components which are then linked together. The first is a model for the yield $Y(t_{init})$ one might expect to realize from a weapon if a spontaneous fission initiates the chain reaction at some time t_{init} after the core first achieves a critical state during its assembly but prior to assembly being completed. The moment at which the criticality parameter α achieves a value of zero is taken to define $t = 0$. (Recall that $\alpha = 0$ for a core of threshold critical mass, and $\alpha > 0$ for a supercritical core.) The second is a probabilistic model for the chance that the reaction will *not* be initiated by time t_{init}; this is based on the material developed in the preceding section. By combining these, one can then make the statement that if P is the probability that a predetonation does *not* occur during the time interval $(0, t_{init})$, then the chance of obtaining *at least* yield $Y(t_{init})$ is $100P$ percent. There are some subtleties to this argument as discussed in what follows, but this is the fundamental idea.

This yield model is adopted from one developed by J. Carson Mark in collaboration with Frank von Hippel and Edwin Lyman (Mark et al. 2009), although the analysis given here is somewhat more general than theirs. In what follows, I refer to their paper as MvHL. As in the previous section, the development here assumes an *untamped* core.

When operation of a weapon is triggered, the core will initially be subcritical, but as it is assembled by either an implosion or a gun mechanism, it will reach a condition where $\alpha = 0$, "first criticality." Subsequently, α will increase until the core is fully assembled. The most desirable situation is that the chain reaction not be initiated until the core reaches its fully-assembled state, as this would result in the most efficient explosion. The value of α in the fully-assembled condition is designated as α_O, and the time at which this happens after first criticality is designated as t_O; see Fig. 4.5. As soon as the chain reaction starts, the core will begin to expand and α will begin to decline. When α reaches zero, "second criticality" occurs, after which the reaction essentially shuts down. α_O is thus the maximum possible value that α can have, and can be thought of as the design or "nominal" value of the weapon's criticality parameter. On the other hand, the *worst* circumstance would be that the reaction is initiated via a spontaneous fission at just the moment when first criticality is achieved. In this case the bomb will (likely) blow itself apart and generate only the minimum possible "fizzle" yield. The important point for the moment, however, is that the yield of a weapon depends essentially on the value of α when the chain reaction begins.

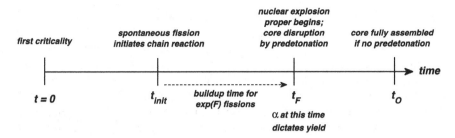

Fig. 4.5 Sketch of time-evolution of predetonation. The core achieves first criticality ($\alpha = 0$) at $t = 0$, and t_O is the time when α reaches its nominal design value α_O. If the chain reaction starts at t_{init}, some time is required for e^F fissions to occur, after which the nuclear explosion proper is underway (see the text)

Here is the main subtlety: This crucial value of α is *not* that at t_{init}, because some additional time is required for the reaction to build up to the point where the pressure exerted by the fission fragments is great enough to begin causing a sensible expansion of the bomb core against any remaining force of the assembly mechanism; that is, the nuclear reaction proper requires some time to become established. Even if the first fission occurs before assembly is complete, it may well be that this build-up time could be great enough to allow completion of the core assembly, resulting in achieving the full design yield. The time after first criticality at which the nuclear reaction proper begins is designated as t_F, so it is the value of $\alpha(t_F)$ that dictates the yield.

In view of the above considerations, we need to pin down three concepts in order to establish an expression for the expected yield. These are (i) A model for the growth of α during the time between first criticality and full assembly; this is needed in order to be able to eventually estimate $\alpha(t_F)$; (ii) A procedure for estimating t_F; and (iii) A model for how the efficiency of the explosion depends on $\alpha(t_F)$. These factors are addressed individually in the following paragraphs.

From Sect. 2.2, the growth rate of the number of neutrons N (or of the neutron density) within a bomb core is described by the differential equation

$$\frac{dN}{dt} = \left(\frac{\alpha}{\tau}\right)N, \tag{4.14}$$

where τ is the mean time that a neutron will travel before causing a fission, ~10^{-8} s.

For the time-dependence of α, I follow MvHL and adopt a simple model: That α grows linearly from zero to α_O over time t_O:

$$\alpha(t) = \left(\frac{\alpha_O}{t_O}\right)t. \tag{4.15}$$

In computing actual numbers, I will take $t_O = 10\,\mu s$ (characteristic of an implosion weapon), but this quantity is left as a general variable in the analysis.

If the reaction starts at some time t_{init} ($\leq t_O$), then the neutron population at some later time t will be given by integrating (4.14) from t_{init} to t after substituting the α-growth model of (4.15) into (4.14). It proves to be more useful, however, to speak in terms of the number of fissions that have occurred between these time limits. Since the number of neutrons created is proportional to the number of fissions that have occurred and since the constant of proportionality will cancel from both sides in (4.14), that equation also dictates the number of fissions that will have taken place. Let the number of fissions that occur over the interval (t_{init}, t) be e^F. Upon integrating we find

$$F = \left(\frac{\alpha_O}{2\,\tau\,t_O}\right)\left(t^2 - t_{init}^2\right). \tag{4.16}$$

This result is central to the developments that follow.

To address point (ii) above, if we had available a value for F_{react} corresponding to the time t_F at which the nuclear explosion can be considered to have become fully established, we could compute that time simply by solving (4.16) for t on setting $F = F_{react}$. To pin this value down, MvHL offer the following argument. Consider a plutonium core of mass 10 kg. The specific heat of Pu is about 130 J/(kg K), and a typical fission releases about 180 MeV of energy. To raise the temperature of the core by a modest 30° would require $F \sim 35$. To melt the core would require $F \sim 38$. Beyond this, to liberate by fissions an amount of energy per gram of material equivalent to that of detonating TNT (~1 kcal/gr) would require $F \sim 42$. But as MvHL point out, by this time the plutonium will have vaporized and begun to exert a pressure on its surroundings in the megabar range, a pressure which will overwhelm any remaining force of the assembly mechanism. Thus, we are justified in taking the nuclear explosion to have started by, say, the time that $F_{react} \sim 45$. This is the value which MvHL adopted, but sensible changes make little difference to the results. F_{react} is left as a general parameter in the development that follows, and is hereafter abbreviated as F.

Now to point (iii). In Sections 17 and 18 of his *Los Alamos Primer*, Serber (1992) develops an argument to show that if α has the value α_F when the nuclear explosion begins (that is, after e^F fission have occurred), then the yield Y of the weapon will behave approximately as.

$$\left(\frac{Y}{Y_O}\right) \sim \left(\frac{\alpha_F}{\alpha_O}\right)^3, \tag{4.17}$$

where Y_O is the nominal design yield. This dependence can be understood from the analysis of efficiency in Sect. 2.5. From (2.96), the efficiency is proportional to $\alpha^2\langle\rho r\rangle$. But numerical solution of the formal criticality conditions with cores of from one to two critical masses shows that $\langle\rho r\rangle$ is roughly linearly proportional to α, which leads to the conclusion that the efficiency must be proportional to α^3. The proportionality for cores of one to two critical masses of ^{235}U is illustrated in Fig. 4.6.

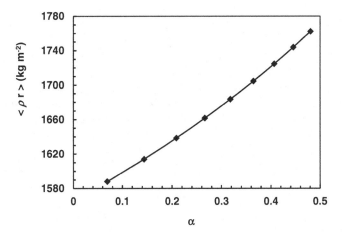

Fig. 4.6 $<\rho r>$ versus α for untamped ^{235}U cores of masses 50, 55, ... 90 kg ($C = 1.09$–1.96). (σ_f, σ_{el}, ν) = (1.235 bn, 4.566 bn, 2.637)

We can now estimate the minimum fractional yield Y/Y_O. The worst-case "fizzle" scenario will be if $t_{init} = 0$. Setting $t_{init} = 0$ in (4.16) and solving for t gives $t_{fizzle} = \sqrt{2t_O \tau F/\alpha_O}$ as the time which must elapse for e^F fissions to have occurred. Substituting this result into (4.15) gives α_F, which, when substituted into (4.17) gives.

$$\left(\frac{Y}{Y_O}\right)_{fizz} = \left(\frac{2\tau F}{\alpha_O t_O}\right)^{3/2}.$$ (4.18)

For (τ, F, α_O, t_O) = (10^{-8} s, 45, 1, 10^{-5} s), $t_{fizz} \sim 3$ μs and $Y/Y_O \sim 0.027$. For a nominal yield of 20 kilotons, this implies a fizzle yield of some 540 tons, entirely ample to destroy a bomb and so alleviate the issue of an adversary being able to recover fissile material. The minimal fizzle yield is now likely a matter of only historical interest, but it could become a very real issue in the event of any forensic analysis in the wake of a terrorist-sponsored weapon.

Even if $t_{init} = 0$, there is still some slight chance of achieving full yield. This will be the case if $t_F > t_O$, for then the core will be able to achieve full assembly before the nuclear reaction proper is underway (also providing that the non-nuclear components of the bomb function properly!) This can be expressed as

$$(t_O)_{full} \leq \frac{2\tau F}{\alpha_O} \quad (t_{init} = 0).$$ (4.19)

For the above values of the parameters, this corresponds to $t_O \leq 0.9$ μs, a tall order.

More generally, if $t_{init} \neq 0$, full yield will be achieved if $t_F > t_O$. Solve (4.16) for t_F and impose this condition; the result is.

$$(t_{init})_{full} \geq t_O \sqrt{1 - \frac{2 \tau F}{\alpha_O t_O}}. \tag{4.20}$$

For the above parameters, this evaluates to about 9.5 μs.

In reality, one will most likely have $(t_O)_{full} < t_{init} < (t_{init})_{full}$, in which case a partial-yield explosion will occur. To examine this situation, take the solution of (4.16) for the time at which e^F fissions have occurred, and use (4.15) and (4.17) to determine the yield:

$$\frac{Y}{Y_O} = \left(\frac{t_{init}}{t_O} \right)^3 \left(1 + \frac{2 \tau t_O F}{\alpha_O t_{init}^2} \right)^{3/2}. \tag{4.21}$$

You should be able to show that this expression gives the correct minimum yield of (4.18) for $t_{init} \rightarrow 0$.

To estimate the probability of achieving the yield predicted by (4.21) requires a model for the spontaneous-fission (SF) predetonation-probability characteristics of the core. As remarked earlier, the premise here is that if P is the probability that a predetonation does *not* occur over the time interval $(0, t_{init})$, then one can state that the chance of obtaining *at least* the yield predicted by (4.21) is $100P$ percent. By investigating the situation for various values of t_{init}, we can build up a plot of the probability of achieving a given fractional yield as a function of fractional yield.

The predetonation model was developed in the preceding section. As applied to the current situation, the average number of SFs during some span of time $(0, t_{init})$ can be written from (4.5) as

$$\mu = M R t_{init}, \tag{4.22}$$

where M is the mass of spontaneously fissioning material and R the rate of SFs. [R is used here to designate the rate of SFs as opposed to F, as the latter symbol is already taken in (4.16). Do not confuse R with the core radius R_{core}, which is introduced below.] If each SF releases on average ν neutrons, the probability of *not* experiencing a predetonation during the selected time interval is, from (4.6) and (4.7),

$$P_{\substack{no \\ predet}} = e^{-\mu} \sum_{k=0}^{} \left(\frac{\mu^k}{k!} \right) \left(P_{escape} \right)^{k\nu}, \tag{4.23}$$

where P_{escape} is the probability that an individual neutron will escape from the core without causing a fission. Reminder: As in Chap. 2, we avoid the issue of a spectrum of neutron-number emission by using an average "effective" value ν.

As described in the preceding section, P_{escape} depends on the maximum number of times S that one is willing to allow neutrons to scatter before escaping the core. The case which results in the highest probability of predetonation (that is, the worst-case estimate) was found to be $S = 0$, and this value is adopted here (Users can change this in the corresponding spreadsheet if desired; see below.) In this case, P_{escape} is directly equal to the spherical-escape probability (4.9) derived in Appendix F:

$$P_{escape} = \frac{3}{8x^3}\left[2x^2 + e^{-2x}(2x + 1) - 1\right], \tag{4.24}$$

where $x = \sigma_{total}\, nR_{core}$.

The spreadsheet **FissionYield.xls** carries out the foregoing calculations. A spherical core is assumed. The user enters values for the mass and density of the core material, the mass of spontaneously-fissioning material, values for the cross-sections and number of neutrons per spontaneous fission, and values for τ, t_O, F, and α_O. Upon entering a value for t_{init}, the non-predetonation probability and fractional yield are calculated. As with the pre-detonation calculations, the spreadsheets take an upper limit of $k = 20$ when computing (4.23). τ, t_O, F and α_O are not used in the calculation of the probability, but are needed to compute the yield according to (4.21).

Figure 4.7 shows results obtained for a 6-kg Pu core contaminated with 1%, 6%, and 20% Pu-240 by mass for $(\tau, F, \alpha_O, t_O) = (10^{-8}$ s, 45, 1, 10 μs$)$. Computations were run for t_{init} from 0.5 to 9.5 μs in steps of 0.5 μs. The probability of achieving full yield with 1% contamination is about 80%; for 6% contamination the full-yield chance falls to about 27%.

The curve for 20% contamination corresponds to what one would expect for reactor-grade plutonium. The chance of achieving any sensible fraction of the design yield with such contamination is abysmal. While this might seem comforting when

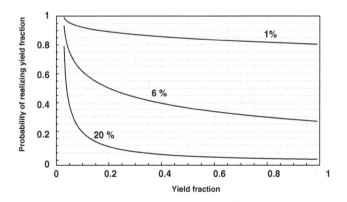

Fig. 4.7 Probability of achieving a given fractional yield as a function of fractional yield for a 6-kg core of ^{239}Pu of normal density (15.6 gr cm^{-3}) contaminated with 1%, 6%, and 20% ^{240}Pu (top to bottom curves). $(\tau, F, \alpha_O, t_O) = (10^{-8}$ s, 45, 1, 10^{-5} s$)$. The values of the nuclear constants for ^{239}Pu are $(\sigma_f, \sigma_{el}) = (1.800$ bn, 4.394 bn$)$; for Pu-240, $(\nu, t_{1/2}) = (2.257, 1.14 \times 10^{11}$ year$)$

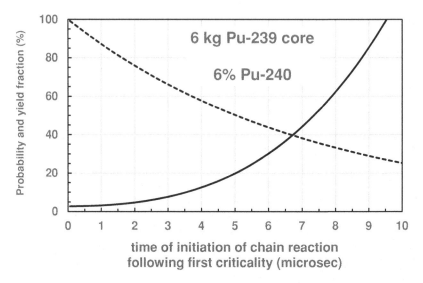

Fig. 4.8 Dashed line: Probability of non-predetonation (or, equivalently, of achieving a given yield) as a function of time of initiation of chain reaction t_{init} following first criticality for a 6-kg core of ^{239}Pu of normal density (15.6 gr cm^{-3}) contaminated with 6% ^{240}Pu. Solid line: Corresponding minimum yield fraction. All other parameters are as in Fig. 4.7

considering the possibility of terrorists trying to develop a Hiroshima or Nagasaki-type bomb based on plutonium extracted from spent fuel rods, bear in mind that a device which realizes even a few percent of its design yield would still create a devastating explosion and disperse radioactive material over a large area.

Figure 4.8 shows another way of displaying some of the information in Fig. 4.7. For fixed values of (τ, F, α_O, t_O), setting the mass, density, and cross sections of the core determine the value of x in (4.24). Specifying the level of spontaneously fissioning contaminant determines the value of μ in (4.22). The common factor between the non-predetonation probability of (4.23) and the minimal yield fraction of (4.21) is the time t_{init} at which the chain reaction begins. We can then plot both of these quantities versus t_{init} for a given core/contaminant scenario. Figure 4.8 shows this for a 6 kg ^{239}Pu core contaminated with 6% ^{240}Pu, which corresponds to the middle curve in Fig. 4.7. The yield curve reaches 100% at $t_{init} = 9.5$ μs, a manifestation of (4.20), but the probability of this occurring is only ~27%. Similarly, the probability of achieving a yield of 50% or better is only about 37%. To achieve this yield requires that the reaction not initiate until at least 7.5 μs elapse after first criticality. For simplicity, these curves were computed using only the first four terms ($k = 0$–3) of (4.23).

4.4 Tolerable Limits for Light-Element Impurities

In addition to the possibility of predetonation caused by spontaneous fission, another danger for weapons designers is that a chain reaction can be initiated by the natural alpha-decay of the fissile material if that material contains even a small percentage of light-element impurities. A particular danger in this regard is the presence of any beryllium in a Pu core. ^{239}Pu has a fairly short half-life for alpha-decay, about 24,100 years, or 7.605×10^{11} s. From the decay-rate formula (4.4), this leads to an enormous rate of alpha-decays:

$$R_\alpha = 10^3 \left(\frac{N_A}{A} \right) \left(\frac{\ln 2}{t_{1/2}} \right) = 10^3 \left(\frac{6.022 \times 10^{23}}{239} \right) \left(\frac{\ln 2}{7.605 \times 10^{11}} \right)$$
$$= 2.296 \times 10^{12} \, \text{kg}^{-1} \, \text{s}^{-1}. \tag{4.25}$$

This figure is much greater than the rate of spontaneous fissions for ^{240}Pu. For a 10-kg core of pure ^{239}Pu, the alpha-decay rate would be 2.3×10^{13} s^{-1}. If some of these alphas should find a beryllium nucleus to react with during the time that the bomb core is being assembled, the result will be a neutron which could go on to initiate a premature detonation via an (α, n) reaction of the sort involved in Chadwick's discovery of the neutron (Sect. 1.4):

$$^4_2\text{He} + ^9_4\text{Be} \rightarrow ^{12}_6\text{C} + ^1_0\text{n}. \tag{4.26}$$

A similar effect happens with alpha bombardment of lithium:

$$^4_2\text{He} + ^7_3\text{Li} \rightarrow ^{10}_5\text{B} + ^1_0\text{n}. \tag{4.27}$$

This is a serious issue. As described by Bernstein (2007), plutonium metal at room temperature is rather brittle and difficult to form into desired shapes unless alloyed with another metal. But a light alloying metal such as aluminum cannot be used because of this (α, n) problem; one needs to use an alloying material whose nuclei have a Coulomb barrier strong enough that they cannot be overcome by alpha-particles of a few MeV. Los Alamos metallurgists alloyed plutonium with gallium to achieve desirable malleability properties.

Chemical processing of plutonium will inevitably introduce some level of impurities. The question is: What level of impurity can one tolerate if the resulting rate of neutron production is to be kept below, say, one per 100 μs? For simplicity, we develop the analysis assuming that only one impurity is present.

To address this issue requires appreciating two empirical ideas from experimental nuclear physics: (i) The *yield* (y) of a reaction; and (ii) The *stopping power* (S) a material presents against particles traveling through it. Note that y here refers to the yield of a particular nuclear *reaction*, not the yield of a *bomb* as whole, for which we have used the symbol Y. We discuss these two issue first, and then develop a formula for predicting the neutron-generation rate for some impurity. For sake of definiteness, I have in mind beryllium as the impurity.

The yield y of a reaction can be understood as follows. Suppose that one has a well-mixed sample of Be and some alpha emitter such as uranium, plutonium, radium, or polonium. Not all of the emitted alphas will find a Be nucleus to react with; atoms are mostly empty space. The yield of the reaction is the number of neutrons produced per alpha emitted. From figures given by Fermi (1950, p. 179), 1 Curie (Ci) of radium well-mixed with beryllium yields about $10-15 \times 10^6$ neutrons per second, whereas 1 Ci of polonium well-mixed with Be yields some 2.8×10^6 neutrons per second. (Both Ra and Po are alpha-emitters.) With 1 Ci $= 3.7 \times 10^{10}$ s^{-1}, these figures correspond to yields of $2.7-4.1 \times 10^{-4}$ and 7.6×10^{-5}, respectively. Radium and polonium alphas have energies of about 4.8 and 5.3 MeV. More energetic alphas actually give lower yield due to the fact that a higher-energy particle will have a longer range of travel in some material before being consumed in a reaction; this is discussed further in the next paragraph. Plutonium alphas have energies of about 5.2 MeV, so we might expect a yield for Pu-alphas on Be somewhere between these two results, say $y \sim 10^{-4}$. This is in the ballpark: West and Sherwood (1982) give the neutron yield of 5.2-MeV alphas on ^{9}Be as 6.47×10^{-5}.

As the terminology suggests, the *stopping power* S of a material is a measure of how effective the material is at stopping particles that are traveling through it. As you might infer, the range of a particle in some material is inversely proportional to its stopping power. Empirically, the *Bragg-Kleeman rule* (Evans 1955, p. 652) states that stopping power is proportional to the mass density of the material and inversely proportional to the square root of its atomic weight:

$$S \propto \frac{\rho}{\sqrt{A}}. \tag{4.28}$$

Suppose that one has a mixture of two materials, A and B, each with their own stopping power for alpha particles, S_A and S_B. If $S_B > S_A$, an alpha will have a greater probability of reacting with a nucleus of material B than one of material A, presumably in the proportion S_B/S_A. We will use stopping power as a measure of relative amounts of "reactivity" of the two materials. For the impurity, the density to be used will not be its "normal" density, but rather that given by its hopefully small mass distributed throughout the volume of the core.

Now consider a bomb core made of a heavy fissile material of atomic weight A_H and density ρ_H along with an admixture of some light-element impurity of atomic weight A_L and density ρ_L as defined above. We presume that the amount of impurity is so slight that ρ_H will be essentially the "normal" density for the core material. Also, let the nuclear number densities of the two materials be n_H and n_L, respectively; the goal here is to get an expression for the tolerable limit on n_L/n_H. If V is the volume of the core, the mass of the impurity will be $n_L A_L V/N_A$, and its mass density will be $n_L A_L/N_A$ (N_A = Avogadro's number). This will give a stopping power S_L according as

$$S_L \propto \frac{\rho}{\sqrt{A}} \propto \frac{n_L A_L}{N_A \sqrt{A_L}} \propto \frac{n_L \sqrt{A_L}}{N_A}, \tag{4.29}$$

and similarly for the heavy fissile material.

The rate of neutron production R_n (neutrons per second) caused by the impurity will depend on the α-decay rate R_α of (4.25) and the yield. If the fissile material itself has no neutron yield for alpha bombardment (Coulomb barrier too great), R_n will be the α-decay rate times the yield, times the fraction of the stopping power due to the imnpurity:

$$R_n = R_\alpha y \left(\begin{array}{c} fraction\ of\ total\ stopping \\ power\ due\ to\ impurity \end{array} \right)$$

$$= R_\alpha y \left(\frac{n_L \sqrt{A_L}}{n_L \sqrt{A_L} + n_H \sqrt{A_H}} \right) \tag{4.30}$$

Unless very poor chemical separation techniques are involved, we should expect $n_L \ll n_H$, so we can simplify this to

$$R_n = R_\alpha y \left(\frac{n_L}{n_H} \right) \left(\sqrt{\frac{A_L}{A_H}} \right). \tag{4.31}$$

For a sepecified tolerable maximum neutron rate R_n, it is more convenient to write this as a constraint on the ratio of the number densities:

$$\left(\frac{n_L}{n_H} \right) < \frac{1}{y} \left(\frac{R_n}{R_\alpha} \right) \sqrt{\frac{A_H}{A_L}}. \tag{4.32}$$

Assuming beryllium as the contaminant in a 10-kg Pu core, adopting the West and Sherwood yield, and taking $R_n = 10^4$ s^{-1} (= one per 100 μs) gives.

$$\left(\frac{n_L}{n_H} \right) < \frac{1}{6.47 \times 10^{-5}} \left(\frac{10^4}{2.3 \times 10^{13}} \right) \sqrt{\frac{239}{9}} \sim 3.5 \times 10^{-5}. \tag{4.33}$$

This means that no more than about 1 atom in 29,000 can be one of beryllium.

In the case of a ^{235}U core the situation is much more forgiving; one can tolerate a very high degree of impurity if necessary. The alpha-decay half-life for ^{235}U is about 7.0×10^8 years, or ~2.2×10^{16} s. This gives $R_\alpha \sim 8.0 \times 10^7$ kg^{-1} s^{-1}, or about 4.0×10^9 s^{-1} for a 50-kg core. For a yield of 5×10^{-5}, (4.32) gives $n_L/n_H < 0.26$.

4.5   Neutron Initiators

Material in this section is adopted from Reed (2019).

An important application of the yield concept introduced in the preceding section involves the question of initiating a nuclear explosion once a core has been assembled and (hopefully) avoided predetonation by spontaneous fission. In the Manhattan Project, this was accomplished by placing a device known as an *initiator* within the core. According to Sublette (2007), this was an approximately golf-ball-sized sphere that contained polonium and beryllium which were initially separated by a metal foil; this is sketched in Fig. 4.9. Upon implosion or by being crushed by the incoming projectile piece, the Po and Be mix; alphas from the Po then strike Be nuclei, liberating neutrons to initiate the detonation. This section explores how the amount of polonium necessary to achieve a given neutron flux can be estimated, and how long it would take to produce such amounts in a Manhattan Project reactor.

To open this analysis, a brief history lesson is appropriate. On June 18, 1943, Los Alamos Laboratory Director Robert Oppenheimer wrote to General Groves regarding anticipated specifications for neutron initiators (Oppenheimer 1943). In part, Oppenheimer's letter read: "The time during which we would like to be certain of a detonation will surely not be less than 10 μs and may … be as much as ten times as long. At this time we should like to have a mean emission of at least 100 neutrons. A source of this strength involves several curies of polonium in adequate contact with beryllium." Oppenheimer's time estimate of 10–100 μs reflects fact that, at the time, both the uranium and plutonium bombs were expected to be of the gun design; the plutonium spontaneous fission crisis, which would lower the requisite timescale to about a single microsecond, had not yet emerged.

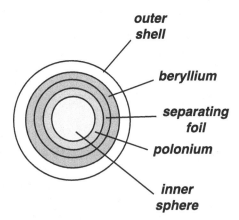

Fig. 4.9 *Conceptual* sketch of an initiator. In this arrangement, a few milligrams of polonium are deposited onto a supporting inner surface. The separating foil, made of a heavy metal such as gold, stops alpha-particles emitted by the Po from striking a surrounding layer of beryllium until the sphere is crushed by the assembling bomb core. In reality, the entire assembly is about an inch in diameter. This sketch is purely schematic and not to scale

Development of the Project's neutron initiators was such a highly-secret affair that it received little attention in many official histories of the work, and so for many years remained overlooked in Manhattan scholarship circles. The polonium program received no mention at all in Henry Smyth's August 1945 "Atomic Energy for Military Purposes" public report on the bomb project (Smyth 1945). There is no index entry for "polonium" or "Dayton" (Ohio, where the work was largely carried out) in either Hewlett and Anderson's 1962 history of the Atomic Energy Commission or Vincent Jones's 1988 history of the Army's role in the Manhattan Project, although Dayton receives a very brief mention in the latter; see (Jones 1988). Manhattan Engineer District material on polonium remained classified until 1983, and some aspects of the associated chemical research are still off-limits to outsiders. The focal point of the polonium work was a highly-secret facility operated by the Monsanto Chemical Company in Dayton, Ohio. The work was overseen by Charles Allen Thomas, Monsanto's vice-president for research. The Dayton work began to receive more recognition in Manhattan Project history circles with the 1993 publication of Hoddeson et al.'s technical history of Los Alamos, and then with the publication of a personal memoir by Sopka and Sopka (2010). But it was not until the 2017 publication of *Polonium in the Playhouse* by Thomas's granddaughter, Linda Thomas, that the Dayton effort begin to receive the recognition it deserves; see (Thomas 2017).

Since the timescale over which an initiator must function may be only a few microseconds and one wants to be sure of having at least dozens of neutrons to absolutely ensure a detonation, the characteristics of the (α, n) source are crucial: One must have a very copious α-emitter. Manhattan Project initiators used polonium-210 (^{210}Po) because of its short half-life (138 days); a mere 0.24 mg emits a full Curie of alpha particles, the same as an entire gram of radium. This short half-life was both a blessing and a curse for Manhattan Project scientists: the half-life guaranteed a steady supply of alpha particles, but, conversely, initiators had very limited shelf-lives once fabricated. Consequently, it became necessary to establish a dependable supply of polonium, which is otherwise a rare element. For practical purposes, there are only two sources: extracting it as a decay product from waste ores from uranium and radium-mining operations, or by breeding it via neutron bombardment of bismuth within a reactor. During the Manhattan Project, some polonium was produced by waste-ores extraction, but the vast majority was synthesized by the bismuth-bombardment process.

The bismuth process begins by irradiating ^{209}Bi (the only stable isotope of that element) by neutrons inside a reactor. Neutron capture transmutes the ^{209}Bi to ^{210}Bi, which subsequently beta-decays to ^{210}Po with a half-life of five days:

$$\ _{0}^{1}n \ + \ _{83}^{209}\text{Bi} \ \rightarrow \ _{83}^{210}\text{Bi} \ \xrightarrow[5.0\,days]{\beta^{-}} \ _{84}^{210}\text{Po}. \tag{4.34}$$

The neutron-capture cross-section for this process is small, however, so it was necessary to irradiate hundreds of pounds of bismuth to produce Curie-level amounts of Po.

For processing in the Hanford reactors, bismuth was formed into five-pound slugs which could be fed into the reactor's fuel tubes. Irradiated slugs were delivered to Dayton by train or truck, guarded by military couriers. By June 1945, Monsanto was shipping 35 Curies of Po per week to Los Alamos; some fifty tons of irradiated bismuth would be processed. The first full-scale initiator was not constructed at Los Alamos until just a few weeks before the *Trinity* test; Hoddeson et al. (1993) relate that the first one was nearly dropped down a drainpipe.

The predetonation issue studied in the preceding two sections gets to the heart of *why* it was necessary to develop initiators. In Sect. 4.2, it was remarked that for a 50 kg *Little Boy* ^{235}U core, only ~3 × 10^{-5} spontaneous fissions will occur over a 100 μs assembly timescale, a number far too small to guarantee detonation when desired. Since initiators had to be developed for the *Little Boy* gun bomb, the thinking at Los Alamos was that they might as well be used in the implosion bomb as well. The calculations which follow are predicated on a neutron-emission time of ~1 μs.

If an initiator is to produce Oppenheimer's 100 neutrons over one microsecond, the number of curies of Po required can be estimated with the yield analysis of the preceding section. Based on the figures given therein drawn from West and Sherwood (1982), I adopt a yield y of polonium alphas on Be of 7 × 10^{-5}. [Specifically, polonium alphas have kinetic energies of ~5.3 MeV. West and Sherwood give the yields of 5.2 and 5.4-MeV alphas on Be as 6.47 and 7.50 × 10^{-5}, so 7 × 10^{-5} seems a sensible compromise.] As in that section, if R_α is the rate of alpha-emissions, then the rate of neutron generation will be $R_n = y R_\alpha$. With $y = 7 \times 10^{-5}$, demanding 100 neutrons per microsecond corresponds to just less than 39 curies of Po, equivalent to about 8.6 mg. This is a substantial amount of any radioactive material; 40 Ci of ^{210}Po would generate a decay heat of some 1.2 Watts.

The analysis of how to arrange for an appropriate rate of production of Po is more complex, and involves two interlinking factors: the instantaneous rate of production within the reactor, and any decay of the polonium so formed before it is extracted from the bismuth. We deal with each of these in turn.

Since the production of polonium depends on having bismuth capture neutrons, the rate of production depends on the available neutron flux within the reactor. For this we can use the neutron flux/reaction rate equation of Sect. 3.3. In that analysis, recall that for a flux of neutrons Φ per unit area per unit time incident on N target nuclei of reaction cross-section σ, the reaction rate will be $R = \Phi N \sigma$ per second. For a reactor generating thermal power P_t Watts, the flux is given by Eq. (3.20):

$$\Phi = \frac{P_t}{E_f N_{235} \sigma_{f5}}, \qquad (4.35)$$

where E_f is the energy liberated per fission reaction in Joules, N_{235} is the number of ^{235}U nuclei present in the fuel, and σ_{f5} is the thermal-neutron fission cross section of ^{235}U.

For the purposes of this section, it is more convenient to work not with the number of nuclei of ^{235}U in the fuel, but rather its mass. The reason for this is that much of the historical documentation refers to production rates per unit operating power per unit mass of fuel. However, the mass of fuel involved is not the mass of ^{235}U alone, but rather that of the entire mass of uranium within the reactor, the majority of which will be ^{238}U. The total number of nuclei is given by this total mass of the fuel, M_{fuel}, divided by its atomic weight A_U of natural uranium, multiplied by Avogadro's number N_A. The number of ^{235}U nuclei is then the total number of nuclei times the fractional abundance f (= 0.0072 for natural uranium) of ^{235}U. In fuel-mass form, (4.35) becomes

$$\Phi = \frac{P_t \, A_U}{E_f \, M_{fuel} \, f \, N_A \, \sigma_{f5}}. \tag{4.36}$$

On adopting $A_U = 0.23803$ kg mol^{-1}, $E_f = 180$ meV, $f = 0.0072$, and $\sigma_{f5} = 585$ b, this reduces to.

$$\Phi = \left(3.254 \times 10^{13} \text{ s}^2 \text{ m}^{-4}\right) \frac{P_t}{M_{fuel}}. \tag{4.37}$$

This expression assumes MKS units: P_t in Watts and M_{fuel} in kg. However, in Manhattan Project documents, fuel loads are usually quoted in US tons and power outputs in megawatts (P_{MW}); also, in the reactor engineering community, neutron fluxes are usually cited in number per square centimeter per second. In these units, (4.37) becomes

$$\Phi = \left(3.587 \times 10^{12} \text{ ton MW}^{-1} \text{cm}^{-2} \text{ s}^{-1}\right) \frac{P_{MW}}{M_{fuel}}. \tag{4.38}$$

In reality, the neutron flux within a reactor is a function of position, with the flux decreasing from the center of the core to the edges. A complete treatment involves the neutron diffusion equation, but a simple rate-equation-based calculation will suffice for order-of-magnitude estimates.

The rate equation is now used a second time, in order to get the rate of production of polonium nuclei. Suppose that a mass M_{Bi} of bismuth of atomic weight A_{Bi} and thermal-neutron capture cross-section σ_{Bi} is introduced into the reactor. The instantaneous rate of production of Po nuclei is then given by.

$$R_{Po} = \Phi \left(\frac{M_{Bi} \, N_A}{A_{Bi}}\right) \sigma_{Bi}. \tag{4.39}$$

Here we have $A_{Bi} = 0.20898$ kg mol^{-1}, and the KAERI website referenced in Appendix B gives $\sigma_{Bi} = 0.03384$ b. These values give.

$$R_{Po} = \left(3.173 \times 10^8 \text{ s}^{-1}\text{kg}^{-1}\right) \frac{P_{MW} \, M_{Bi}}{M_{fuel}}. \tag{4.40}$$

To complicate things further, Manhattan Project documents usually quote amounts of bismuth in US pounds (lb), so in analogy to (4.38) we get

$$R_{Po} = \left(1.587 \times 10^{11} \text{ ton MW}^{-1} \text{ lb}^{-1} \text{sec}^{-1}\right) \frac{P_{MW} M_{Bi}}{M_{fuel}}. \qquad (4.41)$$

However, the rate of production is more complicated than (4.41) lets on. The alpha-decay half-life of Po is so short, about 20 weeks, that loss of Po due to decay must be accounted for if the bismuth remains in the reactor for any appreciable time before being extracted. Since Rutherford and Soddy's discovery of half-life, it has been known that if the number of nuclei at any time is $N(t)$, then the decay rate is given by $\lambda N(t)$, where λ is the decay constant, which can be written in terms of the half-life as $\lambda = (\ln 2)/t_{1/2}$; for Po, $\lambda = 5.80 \times 10^{-8}$ s^{-1}. The *net* rate of creation of Po nuclei is then dictated by the differential equation.

$$\frac{dN_{Po}(t)}{dt} = R_{Po} - \lambda N_{Po}(t). \qquad (4.42)$$

Strictly speaking, the rate of production will decline as bismuth is depleted, but the depletion rate turns out to be extremely miniscule. Another refinement to this model could be to account for the 5-day half-life decay of bismuth to Po; I ignore this time as it is short compared to the timescale over which slugs remained in the piles.

If there is no polonium to begin with, then the solution of (4.42) is $N_{Po}(t) = \left(R_{Po}/\lambda\right)\left(1 - e^{-\lambda t}\right)$, and hence the activity $\lambda N_{Po}(t)$ of the extracted Po after neutron-exposure time t, that is, the number of decays per second, is given by

$$Activity = R_{Po}\left(1 - e^{-\lambda t}\right). \qquad (4.43)$$

Note that as $t \to \infty$ (or, equivalently, $t \gg t_{1/2}$), production and decay activity are in balance. Manhattan Project scientists did not have the luxury of irradiating their bismuth for several half-lives; irradiating for one half-life will capture 50% of the long-term activity.

Results for various scenarios with the Oak Ridge X-10 and Hanford reactors are listed in Table 4.2. The first line gives a prediction made by Oppenheimer in his letter to Groves for the X-10 pile: that irradiating 100 lb of Bi for four months should generate about 9 Ci of Po if the reactor were operated at 20 kW per ton of fuel. The present analysis (second line) indicates about 3.9 Ci and a neutron flux of about 7.2×10^{10} cm^{-2} s^{-1} for these circumstances. A few Curies corresponds to a fraction of a gram, which shows that it is entirely justifiable to neglect any depletion of the bismuth. Oppenheimer's estimate was optimistic, but not wildly so. The third line in the table corresponds to an early-1944 configuration of X-10, which saw a power level of 4 MW achieved with 709 of 1248 fuel channels each loaded with 44 cylindrical uranium fuel slugs of mass about 1.2 kg each (Manhattan District History,

Table 4.2 Polonium production for various X-10 and Hanford scenarios

Scenario	Bismuth (lb)	Power (MW)	Fuel (tons)	Irradiation (days)	Activity (Ci)	Flux (10^{10} cm^{-2} s^{-1})
X-10 Oppenheimer (1943)						
	100	0.02	1	120	9	–
X-10 Oppenheimer (1943)—this analysis						
	100	0.02	1	120	3.9	7.2
X-10 (1944 fuel load)						
	100	4	41.6	120	18.6	35
Hanford Oppenheimer (1943)						
	100	1.25	1	1	4.5	–
Hanford Oppenheimer (1943)—this analysis						
	100	1.25	1	1	2.7	450
Hanford (1945 fuel load)						
	100	250	280	120	173	320

1946). A history of the X-10 pile indicates that it could achieve a neutron flux of ~10^{12} cm^{-2} s^{-1} (likely in the center of the core), so the present estimate of ~3.5 × 10^{11} appears reasonable (Cagle 1953).

The much larger Hanford reactors were designed to operate at 250 MW, and three were built. At the time of Oppenheimer's letter, their power level had not yet been fixed; the fourth line of the Table relates his prediction of realizing 4.5 Ci of Po *per day* for 100 lb of Bi if the piles operated at 1250 kW per ton of fuel. The present analysis (fifth line) gives about 2.7 Ci per day for these numbers, about one-half of his prediction; note the much greater neutron flux than with the X-10 pile. The Hanford piles ultimately operated with full fuel loads of ~280 tons (DOE 2001). As indicated in the last line of the table, the present analysis indicates that 100 lb of Bismuth exposed for 120 days would yield about 173 Ci of Po if the pile operated at 250 MW. With three reactors, this means over 500 Ci every four months, or enough for one 50-Ci initiator about every 12 days—a figure which closely matches the rate of per-bomb plutonium production. It is gratifying to see that the numbers accord so well.

4.6 Estimating the Contribution of ^{238}U to the *Trinity* Yield

It was remarked in Sect. 2.6 that ~30% of the yield of the *Trinity* bomb resulted from fissions of ^{238}U induced by high-energy neutrons in the tamper of that device. This section offers an approximate model for estimating the origin of this figure.

A detailed analysis of the physics of this result would be extremely challenging. Fissions which occur in either the core or the tamper could be caused by neutrons that were originally emitted in either material and subsequently (1) stayed in their birth material and caused a fission, possibly after scattering several times; (2) returned to their birth material after one or more escape/reflection trips into and out of the other material; or (3) caused a fission in the other material, again possibly after several trips into and out of each. The model developed here attempts to incorporate the various possible neutron fates in a plausible way without trying to track multiple generations of neutron creation and travel. To do so, I simplify the situation by assuming that any neutrons which migrate from the core into the tamper remain within the tamper, and that any fissions induced within the tamper arise only from core-escaped neutrons and not from secondary neutrons released in fissions within the tamper. On this basis, it is possible to set up a four-factor formula to model the ratio of ^{238}U to ^{239}Pu fissions,

$$\frac{^{238}\text{U fissions}}{^{239}\text{Pu fissions}} \sim \frac{0.3}{0.7} \sim 0.43. \tag{4.44}$$

Each factor can be estimated from straightforward energy and cross-section considerations, as follows.

Assume that a total of F fissions occur in the plutonium core; the intent here is that this number incorporates fissions due to neutrons arising from anywhere within the core/tamper structure. If the number of secondary neutrons liberated per fission is ν, then the net number of neutrons liberated will be $F(\nu - 1)$; the "–1" appears because one neutron is consumed in causing each fission.

Some secondary neutrons will escape from the core into the surrounding tamper. If the probability of escape is p_{esc}, then $F(\nu - 1)p_{esc}$ neutrons will enter the tamper. However, only those neutrons which are more energetic than the fission threshold of ^{238}U will have a chance of inducing a fission. Designate the fraction of neutrons that are sufficiently energetic as f_{thresh}. This will give $F(\nu - 1)p_{esc}f_{thresh}$ neutrons than can potentially induce fissions in the tamper, but not all will do so because of competing effects, notably inelastic collisions with ^{238}U nuclei. If the probability that a neutron will induce a fission is p_{fiss}, then the four-factor analog of (4.44) is

$$\frac{^{238}\text{U fissions}}{^{239}\text{Pu fissions}} = \frac{F(\nu - 1)p_{esc} \, f_{thresh} \, p_{fiss}}{F}$$
$$= (\nu - 1)p_{esc} \, f_{thresh} \, p_{fiss}. \tag{4.45}$$

Each of these factors is analyzed and estimated in the following paragraphs.

(i) *Secondary neutrons.* This is the most straightforward factor. For ^{239}Pu, $\nu \sim 3.2$, so $(\nu - 1) \sim 2.2$.

(ii) *Escape probability.* The probability that a neutron liberated somewhere within a sphere of radius R will reach the surface of the sphere and escape without being consumed in a fission was analyzed in Sect. 4.2. Applying that model to a 6.3 kg core of ^{239}Pu under a compression ratio of 2.5 gives $p_{esc} = (0.33, 0.48, 0.56, 0.59, 0.61, 0.62)$ for a maximum number of scatterings S from zero to five. For more scatterings, p_{esc} does not change drastically, approaching 0.624 as $S \rightarrow \infty$. For estimating the ^{238}U yield, I adopt $p_{esc} \sim 0.55$.

(iii) The fraction of fission-liberated neutrons above the fission threshold was examined in Sect. 1.8, and estimated as $f_{thresh} \sim 0.55$.

(iv) *Probability of fission.* This is the most difficult of the four parameters in (4.45) to estimate with any precision. Neutrons within the ^{238}U tamper which suffer elastic collisions will not lose much kinetic energy and so can go on to possibly induce a fission, whereas those which suffer inelastic collisions tend to lose so much energy that they fall below the fission threshold. As a simple approach to estimating p_{fiss}, I presume that neutrons which find themselves within the tamper do not escape from the tamper, and take the probability of causing a fission to be the ratio of the sum of the fission and elastic-scattering cross-sections to the total cross-section (fission + elastic scattering + inelastic scattering; see Appendix B):

$$p_{fiss} \sim \frac{\sigma_{fiss} + \sigma_{elastic}}{\sigma_{total}} \sim \frac{5.112 \text{ bn}}{7.707 \text{ bn}} \sim 0.66. \qquad (4.46)$$

Gathering the results of the preceding paragraphs gives

$$\frac{^{238}\text{U fissions}}{^{239}\text{Pu fissions}} \sim (2.2)(0.55)(0.55)(0.66) \sim 0.44. \qquad (4.47)$$

This result is in surprisingly good agreement with the value in (4.44) despite the various approximations involved. The escape probability might well be argued to be somewhat higher than has been estimated by virtue of allowing more scatterings or by adopting a slightly lower compression ratio; conversely, the adopted fission probability is likely optimistic because neutron loss from the tamper has been neglected. The threshold fraction could go either way depending on the actual value of the fission barrier for ^{239}U, but is probably not far wrong. While it is satisfying to see that a simple estimate predicts a result in line with the experimental evidence, this must be regarded as a zeroth-order model which begs for a more substantial analysis.

References

Bernstein, J.: Plutonium: A History of the World's Most Dangerous Element. Joseph Henry Press, Washington (2007)

Bernstein, J.: Nuclear Weapons: What You Need to Know. Cambridge University Press, New York (2008)

Cagle, C.D.: The ORNL Graphite Reactor. Oak Ridge National Laboratory report CF 53-12-126 (1953). https://www.osti.gov/biblio/4375110-oak-ridge-national-laboratory-graphite-reactor

DOE: Historic American Engineering Record: B Reactor (105-B Building), HAER No. WA-164. See particularly, p. 72. https://www.cfo.doe.gov/me70/history/NPSweb/DOE-RL-2001-16.pdf (2001)

Evans, R.D.: The Atomic Nucleus. McGraw-Hill, New York (1955)

Fermi, E.: Nuclear Physics. University of Chicago Press, Chicago (1950)

Fermi, E.: Experimental production of a divergent chain reaction. Am. J. Phys. **20**, 536–558 (1952)

Hewlett, R.G., Anderson Jr., O.E.: A History of the United States Atomic Energy Commission, Vol. 1: The New World, 1939/1946. Pennsylvania State University Press, University Park, PA (1962)

Hoddeson, L., Henriksen, P.W., Meade, R.A., Westfall, C.: Critical Assembly: A Technical History of Los Alamos during the Oppenheimer Years, 1943–1945. Cambridge University Press, Cambridge, UK (1993)

Holden, N.E., Hoffman, D.C.: Spontaneous fission half-lives for ground-state nuclides. Pure Appl. Chem. **72**(8), 1525–1562 (2000)

Hull, M., Bianco, A.: Rider of the Pale Horse: A Memoir of Los Alamos and Beyond. University of New Mexico Press, Albuquerque (2005)

Hyde, E.K.: The Nuclear Properties of the Heavy Elements III. Fission Phenomena, Prentice-Hall, Englewood Cliffs (1964)

Jones, V.C.: United States Army in World War II: Special Studies-Manhattan: The Army and the Atomic Bomb. Center of Military History, United States Army, Washington (1988)

Manhattan Engineer District: Considerable detail on the design and operation of the X-10 pile can be found at https://www.osti.gov/opennet/manhattan_district. See especially, vol. 2, Part II, Book IV

Mark, J.C., von Hippel, F., Lyman, E.: Explosive properties of reactor-grade plutonium. Sci. Global Secur. **17**(2–3), 170–185 (2009)

Oppenheimer, J.R.: Oppenheimer's June 18, 1943, letter to Groves can be found in National Archives and Records Administration, Bush-Conant File Relating to the Development of the Atomic Bomb, 1940–1945, Records Group 227. The microfilm publication number is M1392, and the letter appears on Roll 9. The files can be purchased from the NARA on DVD; on the Roll 9 disk acquired by this author, the letter comprises images 992 and 993.

Reed, B.C.: Predetonation probability of a fission-bomb core. Am. J. Phys. **78**(8), 804–808 (2010)

Reed, B.C.: Fission fizzles: estimating the yield of a predetonated nuclear weapon. Am. J. Phys. **79**(7), 769–773 (2011)

Reed, B.C.: Rousing the dragon: polonium production for neutron generators in the Manhattan Project. Am. J. Phys. **87**(5), 377–383 (2019)

Reed, B.C.: Walther Bothe's graphite: physics, impurities, and blame in the German nuclear program. Ann. Phys. (Berlin) (2020)

Serber, R.: The Los Alamos Primer: The First Lectures on How To Build An Atomic Bomb. University of California Press, Berkeley (1992)

Smyth, H.D.: Atomic Energy for Military Purposes: The Official Report on the Development of the Atomic Bomb under the Auspices of the United States Government, 1940–1945. Princeton University Press, Princeton (1945)

Sopka, K.R., Sopka, E.M.: The bonebrake theological seminary: top-secret Manhattan Project site. Phys. Perspect. **12**(3), 338–349 (2010)

Sublette, C.: Nuclear Weapons Frequently Asked Questions. https://nuclearweaponarchive.org/Nwfaq/Nfaq8.html (2007)

Thomas, L.C.: Polonium in the Playhouse: The Manhattan Project's Secret Chemistry Work in Dayton, Ohio. Trillium, Columbus, OH (2017)

West, D., Sherwood, A.C.: Measurements of thick target (α, n) yields from light elements. Ann. Nucl. Energy **9**, 551–577 (1982)

Chapter 5
Miscellaneous Calculations

In this final chapter, we take up some miscellaneous issues associated with fission weapons. One often reads that a bomb core is warm to the touch; this is investigated in Sect. 5.1. Section 5.2 quantifies just how bright the *Trinity* explosion appeared to the naked eye. Section 5.3 develops a numerical simulation for estimating the production of trace isotopes such as ^{240}Pu in a reactor. Finally, Sect. 5.4 investigates a claim that one sometimes reads: that the energy liberated in an individual fission event is sufficient to make a grain of sand visibly jump.

5.1 How Warm Is It?

Would a plutonium bomb core feel warm to the touch? ^{239}Pu is an alpha emitter with a half-life of 24,100 years. As seen in the preceding chapter, this corresponds to some 2.3×10^{12} alpha-decays per second per kilogram of material. With alphas of energy 5.2 MeV, the power generated by alpha-decay from a 1-kg mass of ^{239}Pu amounts to $P \sim 1.9$ Watts.

We can make a rough estimate of how much hotter such a mass would be than the surrounding air by assuming that this power is dissipated in accordance with a semi-empirical expression known as Newton's Law of Cooling. This expression states that the rate of heat energy loss P (that is, the power emitted) due to convection by a body of surface temperature T to a surrounding environment at ambient temperature T_{amb} is given by

$$P = Ah(T - T_{amb}), \qquad (5.1)$$

where A is the surface area of the body and h is an empirical parameter known as the heat transfer coefficient. The value of h depends on the geometry of the object and the properties of the surrounding environment, which is usually a "fluid" such as air or water. For free convection in steady air, $h \sim 5$–25 W/(m^{2}K).

B. C. Reed, *The Physics of the Manhattan Project*,
https://doi.org/10.1007/978-3-030-61373-0_5

For a 6.3-kg *Trinity*/*Fat Man* core, the alpha-decay rate corresponds to a power output of 12.0 Watts. If spherical, this mass would have a radius of 4.58 cm and a surface area of 2.64×10^{-2} m^2. If we adopt $h = 15$ W/(m^{2}K), then $(T - T_{amb}) \sim 30$ K, that is, the surface of the core will be some 30 K warmer than the surrounding air. The claim of warmth is certainly credible.

This calculation is more than a hypothetical exercise. An experimental technique known as nuclear calorimetry is a non-destructive means to quantify masses of radioactive material by measuring such temperature differences. This technique has applications in areas such as fissile material accounting and safeguards enforcement; see, for example, Bracken and Rudy (2007).

5.2   Brightness of the *Trinity* Explosion

Much of the analysis presented in this section is adopted from Reed (2006).

Nuclear weapons release fantastic amounts of energy, only some fraction of which is in the form of visible light. However, they rapidly ionize and heat the surrounding air to incandescence, creating extremely bright fireballs. Rhodes (1986, p. 672) remarks of the July 16, 1945, *Trinity* test that "Had astronomers been watching they could have seen it reflected from the moon, literal moonshine," an allusion to Ernest Rutherford's famous dismissal of the prospects for atomic energy. Investigating this impressive claim makes for an informative exercise in the physics of astronomical magnitudes, and prompts other questions: What fraction of the bomb's yield was in the form of visible light? How bright would the explosion have appeared to an observer on the moon? What about an observer on Mars or otherwise located in the solar system?

These questions can be addressed with the help of information published in a report on the *Trinity* test prepared by the test's director, Kenneth Bainbridge. His report, titled *Trinity*, was prepared soon after the test as Los Alamos report LA-1012. In 1976, a public version of this report was released as Los Alamos report LA-6300-H, and is freely available from the Federation of American Scientists (FAS) website at http://www.fas.org/sgp/othergov/doe/lanl/docs1/00317133.pdf. On page 52 of this report appears a graph of the illumination created by the *Trinity* test in "Suns" equivalent as a function of time at a detector located 10,000 yards from the explosion; this is reproduced in Fig. 5.1. At $t = 10^{-4}$ s the illumination was approximately 80 Suns; it dropped to about 0.1 Suns at $t \sim 0.04$ s, rose back to about 2 Suns at $t = 0.4$ s, and then declined to about 0.4 Suns at $t \sim 10$ s. This "double maximum" in the time-evolution of visible radiation is uniquely characteristic of a nuclear explosion; the reason for this is described later in this section.

In working the following analysis, it must be remembered that many *Trinity* diagnostic experiments were overwhelmed by the explosion, and so yielded only approximate results; the following calculations should be regarded as estimates at best. I interpret "Suns" of illumination to mean multiples of the so-called solar constant, the measured value of the flux of solar energy at the Earth, about 1400 W/m^2.

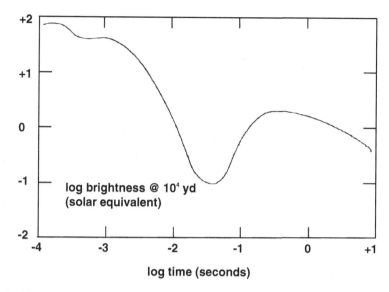

log time (seconds)

Fig. 5.1 Brightness of the *Trinity* explosion as a function of time. Scales are logarithmic. The quality of the curve is somewhat erratic as this graph was produced by scanning a copy of Fig. 7 of Los Alamos report LA-6300; the original version contains numerous grid lines which have not been reproduced to avoid cluttering the diagram

In order to determine the brightness of the *Trinity* explosion as it would have been seen from various vantage points, it is most convenient to work with its equivalent astronomical magnitude. For readers not familiar with the magnitude scale, details can be found in any good college-level astronomy text; the relevant relationships are briefly summarized here without extensive derivation.

For various historical, physical, and physiological reasons, the scale of astronomical magnitudes is defined in terms of the common logarithm of the measured brightnesses of stars. The apparent magnitude m of a star is defined in terms of its measured brightness b (its power flux in Watt m^{-2}), such that the difference between the apparent magnitudes of two stars A and B is given by

$$m_A - m_B = 2.5 \log(b_B / b_A). \tag{5.2}$$

In practice, this relationship is applied to a given star by measuring its brightness in comparison to that of a "standard" star using the same telescope and instrument; the standard star is assigned an arbitrary apparent magnitude. Historically, the star Vega was taken to define $m = 0$.

The absolute magnitude M of a star is defined in analogy to (5.2) but with the measured brightnesses replaced by the true energy outputs of the stars in Watts. In astronomical parlance, energy outputs are known as luminosities, and are traditionally designated by the symbol L:

$$M_A - M_B = 2.5 \log(L_B/L_A). \tag{5.3}$$

The apparent and absolute magnitudes of a star are related via its distance; the inverse-square law of light leads to the relationship

$$m - M = 5 \log(d_{pc}) - 5, \tag{5.4}$$

where d_{pc} designates the distance of the star in parsecs (pc). By definition, the apparent and absolute magnitudes are equal for a star at a distance of 10 pc. One parsec is defined as the distance a star must be from the Sun in order that it has a parallax of one second of arc when viewed from a baseline equal in length to the Earth's orbital radius of one Astronomical Unit (AU). One parsec is equivalent to $206265\ \mathrm{AU} = 3.086 \times 10^{16}\ \mathrm{m} = 3.26$ light-years. The closest star to the Sun, Proxima Centauri, is about 1.3 pc (4.2 light-years) distant.

Equations (5.2)–(5.4) reflect the historical definition of astronomical magnitude as originally developed by Hipparchus around the second century B.C, who defined the brightest stars visible to the naked eye to have $m = +1$ and the faintest as having $m = +6$. Numerically *lower* magnitudes are associated with *brighter* objects. In the modern calibration of magnitudes, Sirius has $m \sim -1.4$, whereas Venus, at its brightest, appears at $m \sim -4.5$. The full moon has $m \sim -12.7$ and the Sun $m \sim -27$.

We now apply these concepts to the *Trinity* (*TR*) explosion by comparing it to the Sun (*S*). From (5.3), the absolute magnitudes of these two sources of illumination are related to their luminosities according as

$$M_{TR} = M_S + 2.5 \log(L_S/L_{TR}). \tag{5.5}$$

Now, let N represent the equivalent number of Suns of *Trinity* illumination incident at some moment on a detector at a distance of 10,000 yards from the explosion. Define the solar constant to be C. For a spherically symmetric explosion, *Trinity's* total power (in Watts) will be $L_{TR} = 4\pi r^2 CN$, where r designates 10,000 yards. Hence

$$M_{TR} = M_S + 2.5 \log(L_S/4\pi r^2 CN). \tag{5.6}$$

The measured absolute magnitude and luminosity of the Sun are +4.82 and 3.83 $\times\ 10^{26}$ W, respectively. (Strictly, these numbers apply for light emitted in the visible part of the electromagnetic spectrum). Setting $C = 1400\ \mathrm{W/m^2}$ and $r = 10{,}000$ yd $= 9144$ m, (5.6) gives

$$M_{TR} = 40.86 - 2.5 \log(N). \tag{5.7}$$

By combining (5.4) and (5.7), we can derive an expression for the apparent magnitude of *Trinity* as viewed from distance d_{pc} parsecs:

$$m_{TR} = 35.86 + 5 \log(d_{pc}) - 2.5 \log(N). \tag{5.8}$$

For practical purposes, solar-system distances are more conveniently measured in AUs: $d_{pc} = d_{AU}/206265$. With this conversion, (5.8) becomes

$$m_{TR} = 9.29 + 5\log(d_{AU}) - 2.5\log(N). \tag{5.9}$$

We are now ready to compute *Trinity* apparent magnitudes. Consider first an observer located on the moon, with $d = 384{,}400$ km $= 2.57 \times 10^{-3}$ AU. With $N = 80$, we find $m_{TR} = -8.4$. Neglecting any effects due to atmospheric absorption and cloud cover, *Trinity* would momentarily have appeared *over 30 times brighter than Venus* to an observer located on the moon; apply (5.2) in the sense of comparing the brightnesses of the two. Not until the fireball cooled to $N \sim 2.2$ a few tenths of a second after the explosion would it have diminished to the brightness of Venus for our lunar observer, and, even after 10 s ($N \sim 0.4$, $m \sim -2.7$) would still have outshone Jupiter ($m \sim -2$). In actuality on the morning of the *Trinity* test, the moon was at first-quarter phase and had set about 1 AM New Mexico time, some four and one-half hours before the test.

Figure 5.2 shows curves of *Trinity* apparent magnitude as a function of distance in AUs for various values of N. At the time of the test, Mercury, Venus, and Mars were respectively 0.97, 0.88, and 1.65 AUs from the Earth. When $N = 80$, *Trinity*

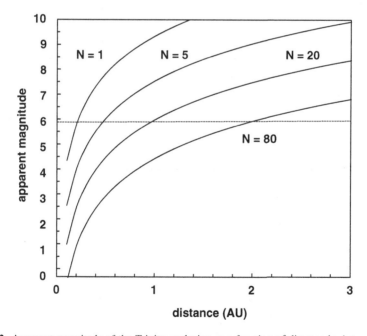

Fig. 5.2 Apparent magnitude of the Trinity explosion as a function of distance in Astronomical Units for times when the explosion was equivalent to 80, 20, 5, and 1 times the solar illumination for a detector located at 10,000 yards. Apparent magnitudes below the dashed line are visible to the naked eye. On this scale, the moon would be located at the extreme left edge of the plot

would have appeared brighter than $m = +6$ for observers residing on all three of these planets, although only Venus and Mars were above the horizon at the time.

Could astronomers have detected the light of *Trinity* as reflected from the Moon? The Moon is seen by reflected sunlight, so the issue is how the flux of *Trinity* light at the moon would have compared with that from the Sun. In his LA-6300 report, Bainbridge remarks that the total radiant energy density received at 10,000 yards was 12,000 J m^{-2}. If we presume that all of this light was emitted over one microsecond, such an energy density corresponds to a flux of 6.8 W m^{-2} at the distance of the moon. The solar flux at the moon will be essentially the same as that at the Earth, about 1400 W m^{-2}, some 200 times greater. A change of one part in 200 corresponds to ~0.005 magnitudes, which would have been difficult to detect with 1945-era observatory technology. The idea of reflected *Trinity* light being visible from Earth is literary license.

Finally, we can estimate what fraction of *Trinity's* yield was in the form of visible light. Various estimates of the yield can be found in the literature; I use 20 kt TNT equivalent. Explosion of one ton of TNT liberates 4.2×10^9 J of energy; 20 kt would be equivalent to 8.4×10^{13} J. If the energy of the explosion radiated uniformly in all directions, an energy density of 12,000 J m^{-2} at 10,000 yards corresponds to a total energy of 1.26×10^{13} J, or approximately 15% of the 20-kt total. The fraction of a bomb's energy emitted as thermal radiation was cited as ~35% in Sect. 2.8.3; "thermal" would include photons over all wavelengths, so we should expect the visible fraction to be smaller than this, as has been estimated here. Well over half of *Trinity's* energy release was in forms *in*visible to the human eye.

These brightness calculations have an interesting historical connection. In the published version of the *Los Alamos Primer*, Robert Serber admits that he overlooked the brilliance of the fireball as a potential source of damage (Serber 1992). Anybody who has stood outside under the noonday Sun on a hot summer day knows how little exposure is required to get a serious sunburn. A nuclear weapon briefly acts like a small Sun, leading to what are known as "flash burns."

Serber offered some figures of his own for estimating brightness. He adopted the radius of the fireball as being about 425 feet (130 m) at about three-tenths of a second after the explosion, at which time he estimated that it would have a temperature of about 7,000 C (~7,270 K). [At $t = 0.3$, Eq. (2.130), drawn from Geoffrey Taylor's analysis of the *Trinity* explosion indicates a radius of ~360 m, but this will not affect the calculations which follow, as the specific time is not involved.] Serber then estimates that for an observer one mile away (1600 m), the fireball would cover an area of the sky about 350 times as large as the Sun, and be about 3.5 times as bright. As argued in what follows, his size estimate is accurate, but his brightness estimate seems much too low.

The apparent size of the fireball can be estimated with trigonometry. For an observer 1600 m away, the angular diameter of the fireball would be about $(260/1600) = 0.16$ radians, or about 9.3°. The angular diameter of the Sun is about 0.5°, so the fireball would appear about 18.6 times wider and thus cover an area $18.6^2 \sim 345$ times that of the Sun, as Serber claimed.

To estimate brightness we can use the Stefan-Boltzmann law, which says that the total radiant power output of an object of radius R and absolute temperature T is proportional to $R^2 T^4$. If d is the distance of the observer from the fireball, the brightness will be proportional to $R^2 T^4/d^2$. The brightnesses of the fireball (F) and Sun (S) will then compare as

$$\frac{b_F}{b_S} = \left(\frac{R_F}{R_S}\right)^2 \left(\frac{T_F}{T_S}\right)^4 \left(\frac{d_S}{d_F}\right)^2. \tag{5.10}$$

The Sun has $R_S = 6.96 \times 10^8$ m, effective surface temperature $T_S \sim 5{,}770$ K, and is about 1.5×10^{11} m distant from Earth. For the 130-meter radius fireball with $T_F = 7{,}270$ K observed from a distance of 1600 m,

$$\frac{b_F}{b_S} = \left(\frac{130}{6.96 \times 10^8}\right)^2 \left(\frac{7270}{5770}\right)^4 \left(\frac{1.5 \times 10^{11}}{1600}\right)^2 \sim 770. \tag{5.11}$$

This value accords very roughly with what can be inferred from Fig. 5.1. At $t \sim 0.3$, the brightness at 10,000 yards $= 9144$ m was $\sim 10^{0.25} \sim 18$ Suns. For an observed at 1600 m, this corresponds to $\sim 18(9144/1600)^2 \sim 590$ Suns. Equation (5.11) is approximate as not all of the power is radiated in the visible part of the spectrum, but since the temperatures are similar the relevant coefficients would largely cancel when computing the ratio of the brightnesses.

The fireball will briefly exist as an apparently large, incredibly brilliant source of light and heat; the resulting exposure will literally vaporize nearby observers and ignite fires to great distances. At Hiroshima, people suffered burns to distances of 7,500 feet from ground zero, roof tiles were melted to 4,000 feet, telephone poles were charred to 9,500 feet, and fires were started to about 15,000 feet. At Nagasaki, fire damage extended to up to 10,000 feet until it was stopped by a river.

Why does Fig. 5.1 exhibit a double maximum? Much of the immediate energy from a nuclear explosion is in the form of X-rays and ultraviolet light, and since cold air is opaque to radiation at these wavelengths, the air surrounding the weapon absorbs the energy and heats up dramatically, to a temperature of about 1,000,000° out to a radius of a few feet. Because this bubble of hot, incandescent air emits energy in the X-ray and ultraviolet regions of the electromagnetic spectrum, it will be *invisible* to an outside observer. But the bubble is surrounded by a cooler envelope, which, although incredibly hot by everyday standards, will be visible to observers at a distance. The temperature of this surrounding air, however, has little physical significance as far as measuring the energy release of the bomb is concerned. As the fireball increases in size, its total light emission increases, up to a first maximum (Stefan's Law: Emission is proportional to surface area times the fourth power of the temperature), after which it begins cooling due to the growing mass of accreted air. Like a hot-air balloon, the fireball will also rise. The temperature within the fireball is so great that all of the weapon residues will be in the form of vapor, including the fission products. As the fireball expands and cools, these vapors condense to form a cloud of solid debris particles; the fireball also picks up water from the atmosphere. All of this material will eventually become fallout, sometimes in the form of radioactive rain. As the fireball ascends, cooling of its outside and air drag often creates a toroidal (doughnut-like) shape. At this stage, the cloud will often have a reddish appearance due to the presence of nitrogen-oxide compounds at its surface.

The air inside the fireball cools by successive radiation and re-absorption of X-rays. When the air has cooled to a temperature of about 300,000°, a "hydrodynamic shock" forms, a so-called "front" of compressed air. The shock front travels faster than energy can be transported by successive absorption and re-emission of radiation, so it "decouples" from the hot sphere and moves out ahead of the latter, leaving behind a region of relatively cool air which "eats into" the central hot sphere. For outside observers, visible radiation comes from the shock wave. As the shock front cools, its observable temperature bottoms out at a minimum of about 2,000°. The shock front also becomes transparent; an observer, if he or she still has eyes, can now look into higher-temperature air, which results in a second brightness maximum. During this time, however, the central fireball is still hot enough to be essentially opaque, and hence invisible.

The brightness of the *Trinity* test was impressive, but perhaps even more staggering was the amount of radioactivity generated: An estimated *one trillion* Curies (see Exercise 5.1). However, nuclear weapons derive their military value not from their radioactivity, but rather from blast and burn effects; militarily, the radioactivity can be a nuisance if you want to move your own troops into the bombarded area. Many of the monitoring instruments deployed for the *Trinity* test were destroyed by the blast, but one pressure gauge located at 208 feet from the base of the 100-foot high tower atop which the device was mounted gave a reading of about 5 tons per square inch, or nearly 700 atmospheres. On considering that reinforced multistory buildings will be demolished by a pressure excess of 20 lb per square inch (~1.4 atmospheres), one can get a sense of the immense destruction caused by nuclear weapons; at Nagasaki, an area of about three square miles was essentially totally destroyed. The situation becomes even more sobering when one learns that postwar improvements resulted in drastically higher weapon yields. The largest pure fission weapon ever detonated by the United States, the *Ivy King* test of November, 1952, generated a yield of 500 kt. This, however, paled in comparison to the February, 1954, *Castle Bravo* test of a "thermonuclear" device which yielded 15 *megatons*. Such fusion-based "hydrogen bombs" use fission bombs as triggering mechanisms. At the opposite extreme, the M-28 "Davy Crockett" nuclear device could be fired from a tripod-mounted recoilless rifle in battlefield conditions; its yield was on the order of "only" 10–20 tons TNT equivalent.

A side-effect of nuclear weapons which can act against friendly forces as well as an adversary is the so-called "electromagnetic pulse," which arises from gamma-rays emitted by the explosion ionizing the surrounding air. Negatively-charged electrons move outward more rapidly than the much heavier positively-charged ions, leading to a strong and rapidly time-varying electric field which can induce damaging currents in electronic equipment. In one spectacular case, detonation of a 1.4-megaton weapon at an altitude of 400 km near Johnston Island in the Pacific Ocean in 1962 caused street lights to fail in Hawaii, some 1400 km away. Readers interested in exploring the effects of nuclear weapons are encouraged to consult the extensive and authoritative volume prepared by Glasstone and Dolan (1977).

5.3 A Model for Trace Isotope Production in a Reactor

In Sect. 3.3, we examined the production of ^{239}Pu in a reactor via calculations that involved only the isotopes ^{235}U, ^{238}U, and ^{239}Pu; no account was taken of other isotopes that are produced along with ^{239}Pu. In Sects. 4.2 and 4.3, however, we saw that even a small amount of ^{240}Pu in a bomb core can lead to significant predetonation issues because of its high spontaneous fission rate. It was remarked in that section that formation of ^{240}Pu in a reactor is inevitable on account of neutron capture by already-synthesized nuclei of ^{239}Pu. In this section, a numerical simulation is developed to approximately quantify the rate of production of ^{240}Pu.

The idea here is to simulate the time-evolution of the abundances of a few key isotopes in a reactor of given thermal power output and fuel load. Reactor engineering is an extremely complex discipline, so a number of simplifying assumptions have to be made for the purpose of a pedagogical model. The simulation is encoded in the spreadsheet **Reactor.xls**.

In developing any reactor simulation, the first issue to decide is what isotopes are to be tracked. Figure 5.3 flowcharts reactions considered in the present case. ^{236}U can be formed from neutron capture by ^{235}U. ^{236}U has a small thermal neutron-capture cross-section of its own, but as this is only about 5 barns it is neglected; ^{236}U is assumed to accumulate as an end product. As in Sect. 3.3, I assume that the creation of ^{239}Pu via neutron capture by ^{238}U is an instantaneous process; no account is taken of the intermediate ^{239}U and ^{239}Np nuclei. The neutron capture cross-sections for ^{239}Pu, ^{240}Pu, and ^{241}Pu are all fairly large, so these species are tracked; ^{242}Pu is assumed to accumulate like ^{236}U as an end product. ^{235}U, ^{239}Pu, and ^{241}Pu all have appreciable fission cross-sections, so those processes must be tracked as well; of course, the vast majority of the energy generated comes from fission of ^{235}U.

The simulation is predicated on a constant number of atoms within the reactor's fuel load, so I assume that when a nucleus fissions, it gives rise to a single nucleus of "fission product." The simulation is programmed to track two fission products should the user desire, with provision for assigning a neutron-capture cross section for the generation of fission product "2" from fission product "1". The results discussed below assumed zero-cross section for this process, but this can easily be changed by the user. In reality, most fission products have half-lives of but a few hours and so decay quickly.

I also assume that no fresh fuel is loaded into the reactor during the span of the simulation. Some smaller cross-sections, such as that for fission of ^{238}U, are neglected, and no decay processes are presumed to occur.

To formulate the simulation, we can begin with the logic advanced in Sect. 3.3: That if P_t is the *thermal* power generated by the reactor and E_f is the average energy

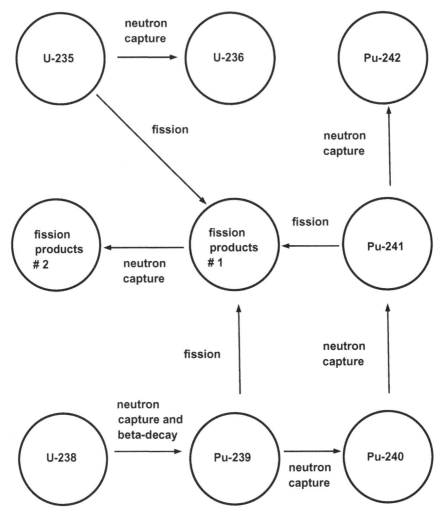

Fig. 5.3 Flowchart for species tracked in reactor simulation. The relevant cross-sections appear in Table 5.1

per fission, then the rate R of fissions must be $R = P_t/E_f$. Converting to power in megawatts, E_f in MeV, and a time unit of one day, the daily rate of fissions must be

$$fissions/day = R = \left[\frac{(86400)(10^6) P_t}{1.6022 \times 10^{-13} E_f} \right]. \tag{5.12}$$

The unit of time in the simulation is taken to be one day, but the user can set the timestep Δt in days or fractions thereof as desired. Over such a timestep, we will have

$$fissions \, per \, \Delta t \, days = R \, \Delta t. \tag{5.13}$$

As in Sect. 3.3, let v be the number of neutrons released per fission. As described below, a consistency requirement demands that v be a function of time in the simulation. The number of neutrons released over time Δt will then be

$$neutrons \, released \, per \, \Delta t \, days = v \, R \, \Delta t. \tag{5.14}$$

The simulation operates by tracking the fractional abundances of isotopes as a function of time. For a given isotope i, let $F^i(t)$ be the fractional abundance of that isotope in the fuel at time t. F is used here for fractional abundance as opposed to the f of Sect. 4.1, as the latter is used here to represent fission. If N is the total number of atoms of fuel loaded into the reactor, then the number of atoms of isotope i at time t will be $N^i(t) = N \, F^i(t)$.

Now consider some process p that a nucleus can suffer under neutron bombardment; this will be either fission (f) by or capture (c) of the neutron. No other processes are allowed to occur, and all free neutrons are assumed to either cause a fission or to be captured during a given timestep. The total cross-section available for all processes over all isotopes at any time is given by the abundance-weighted sums of all individual-process cross-sections in play:

$$\sigma_{total}(t) = \sum_{i,p} \sigma_p^i \, N^i(t)$$
$$= N \left\{ F^{235} \left(\sigma_f^{235} + \sigma_c^{235} \right) + F^{238} \left(\sigma_c^{238} \right) + F^{239} \left(\sigma_f^{239} + \sigma_c^{239} \right) \right.$$
$$\left. + F^{240} \left(\sigma_c^{240} \right) + F^{241} \left(\sigma_f^{241} + \sigma_c^{241} \right) + F^{Prod-1} \left(\sigma_c^{Prod-1} \right) \right\}. \tag{5.15}$$

The total fission cross-section at any time will be

$$\sigma_{fiss}(t) = N \left\{ F^{235} \left(\sigma_f^{235} \right) + F^{239} \left(\sigma_f^{239} \right) + F^{241} \left(\sigma_f^{241} \right) \right\}. \tag{5.16}$$

This next step will require some reflection. From (5.14), we know that the number of neutrons released over Δt days will be $vR \, \Delta t$. The probability that each of these neutrons will go on to cause other fissions will be $(\sigma_{fiss}/\sigma_{total})$, which is given by (5.16) divided by (5.15). Hence we can claim that

$$\left(\begin{array}{c} subsequent \, fissions \, caused \, by \\ neutrons \, released \, over \, \Delta t \, days \end{array} \right) = v \, R \, \Delta t \left(\frac{\sigma_{fiss}}{\sigma_{total}} \right). \tag{5.17}$$

Now, if the power output of the reactor is to remain steady, this subsequent number of fissions must just equal $R\Delta t$ of (5.13), which sets a constraint on v:

$$v = \left(\frac{\sigma_{total}}{\sigma_{fiss}} \right). \tag{5.18}$$

Since the cross-sections will vary in time due to the varying abundance fractions, v must also vary; it has to be computed afresh at each timestep. v will increase with time since the fission cross-section will decrease due to consumption of ^{235}U.

The next step is to set up expressions for how many nuclei of isotope *iso* undergo a given process during time Δt. Nuclei of a given isotope can be created from neutron capture by a nucleus of *lower* weight (if applicable), while simultaneously being lost due to fission and capturing neutrons themselves to produce fission products or isotopes of greater weight. Now, from above, the number of neutrons released over Δt days will be $vR\,\Delta t$. If all of these neutrons are involved in some event over time Δt, then the number of events that correspond to some process will be given by the total number of events involved multiplied by the ratio of the total cross-section for that process to the total available cross-section. Hence, for a given isotope we can write

$$N^{iso}(t + \Delta t) = N^{iso}(t) + \left(\frac{v\,R\,\Delta t}{\sigma_{total}} \right) N \left[F^{lower}\sigma_c^{lower} - F^{iso}\sigma_f^{iso} - F^{iso}\sigma_c^{iso} \right].$$

(5.19)

The simulation actually tracks fractional abundances, that is, (5.19) divided by N:

$$F^{iso}(t + \Delta t) = F^{iso}(t) + \left(\frac{v\,R\,\Delta t}{\sigma_{total}} \right) \left[F^{lower}\sigma_c^{lower} - F^{iso}\sigma_f^{iso} - F^i\sigma_c^{iso} \right].$$

(5.20)

We will need to know N, however, as it remains in σ_{total} through (5.15). For simplicity in determining N, it will be assumed that the fuel is initially composed entirely of ^{238}U. Given that the fuel in most reactors is enriched to only a few percent ^{235}U, this will not be a drastic approximation.

What of the fission products? "Product 1" accumulates from fissions of the three fissile isotopes in the simulation, ^{235}U, ^{239}Pu, and ^{241}Pu, but is lost according as its own abundance and capture cross-section for neutrons. In terms of fractional abundances,

$$F^{Prod-1}(t + \Delta t) = F^{Prod-1}(t)$$
$$+ \left(\frac{v\,R\,\Delta t}{\sigma_{total}} \right) \left[\begin{array}{c} F^{235}\sigma_f^{235} + F^{239}\sigma_f^{239} \\ +F^{241}\sigma_f^{241} - F^{Prod-1}\sigma_c^{Prod-1} \end{array} \right]. \quad (5.21)$$

Similarly, product 2 accumulates from neutron capture by product 1; there is no loss mechanism for product 2:

$$F^{Prod-2}(t + \Delta t) = F^{Prod-2}(t) + \left(\frac{v\,R\,\Delta t}{\sigma_{total}} \right) \left[F^{Prod-1}\sigma_c^{Prod-1} \right]. \quad (5.22)$$

Table 5.1 Adopted thermal-neutron cross-sections (bn) for reactor simulation

Isotope	Fission	Capture
^{235}U	585	99
^{236}U	0	0
^{238}U	0	2.68
^{239}Pu	750	271
^{240}Pu	0	290
^{241}Pu	1010	361
^{242}Pu	0	0
Product 1	0	0

To run the simulation, the user sets the cross-sections and initial abundance fractions of ^{235}U and ^{238}U at $t = 0$. The user also specifies the desired power output, the timestep Δt, and the total mass of fuel in kg. Initial values for the total and fission cross-sections and ν are computed from (5.15), (5.16), and (5.18). New fractional abundances for each isotope and the fission products at time $t + \Delta t$ are then computed according as (5.20)–(5.22). The cross-sections and ν are updated, and the process iterated. Rows of the spreadsheet correspond to timesteps (250 altogether), while columns hold abundances for the various isotopes. For practical purposes, it makes sense to run the simulation only to a time such that ν remains less than the maximum value it could attain in reality, $\nu \sim 2.5$.

Relevant cross-sections are collected in Table 5.1.

The simulation also tracks what is known to reactor engineers as the "burnup" or "fuel exposure" in megawatt-days per metric ton (MWd/MT). This is the cumulative amount of thermal energy produced by the reactor per metric ton of fuel. One metric ton is 1000 kg, and a megawatt-day literally means one megawatt times one day: $(1.0 \times 10^6$ J/s) $(86,400$ s) $= 8.64 \times 10^{10}$ J. According to Mark (1993), a burnup of 33,000 MWd/MT $(= 33$ GWd/MT) is characteristic of commercial reactors.

We now apply this model to the Hanford reactors of the Manhattan Project. According to a Department of Energy publication (DOE 2001), these reactors were each fueled by feeding slugs of natural uranium (hence $^{235}F = 0.0072$, initially) through 2004 aluminum "process tubes" that passed through the piles. During normal operation, each tube contained 32 slugs each measuring 1.44 inches in outside diameter by 8.7 inches long. At a density of 18.95 gr cm^{-3}, this would correspond to just under 4.4 kg per slug, or a total fuel load of about 282,000 kg. I round this to 275 MT for computational purposes as the slugs were jacketed in a thin layer of aluminum. The reactors operated at a thermal power output of 250 MW, and a given slug was irradiated for typically 100 days before being removed and processed to extract its resident synthesized plutonium.

Assuming 180 MeV per fission, the simulation shows that after 100 days, a total of 18.57 kg of plutonium will have been produced, of which 99.66% is ^{239}Pu and 0.34% is ^{240}Pu. This overall plutonium production rate agrees closely with that estimated in Sect. 3.3, 190 gr/day. The burnup to 100 days is about 91 MWd/MT. The initial value of ν is 1.801, which rises to 1.807 after 100 days. In Sect. 4.2, the ^{240}Pu contamination fraction for the *Trinity* and *Fat Man* devices was assumed to be 1.2%, but it must be remembered that the simulation developed here does not account for all processes going on within the reactor. If the *Fat Man* predetonation probability calculation of that section is repeated for a 6.3 kg core containing 0.34% ^{240}Pu (0.0211 kg), the probability that the bomb will function correctly for a 100 μs assembly time is only 45% (zero scatterings); implosion would still be required to achieve a sensibly high probability of avoiding predetonation.

5.4   Can Fission Make a Grain of Sand Visibly Jump?

The idea of a grain of sand being propelled off your desk by the fission of a single uranium nucleus is an appealing image, but, alas, the physics does not hold up. I do not know where this claim originated, but it can be found in, for example, Gerard DeGroot's book *The Bomb: A Life*, where he mistakenly attributes it to having been calculated by Lise Meitner and Otto Frisch when they were preparing their paper on the physics of fission [DeGroot (2004)]. I analyzed the physics of this claim in a paper published in *The Physics Teacher* [Reed (2018)].

Some numbers: The most common form of sand is silicon dioxide, usually in the form of quartz, which has a density of 2.65 g cm^{-3}. A grain of diameter 1 mm will have a mass of about 1.4 mg. The energy released in fission averages about 170 million electron-volts (MeV) per event, or about 2.7×10^{-11} J. If all of this energy is directed into projecting the grain upwards, then the usual *mgh* formula for potential energy shows that we can expect to reach a maximum height of about 0.002 mm, or a mere 1/250 of the radius of the grain itself. Unless the grain is much smaller than this or you have super-power eyes, you are unlikely to be able to discern this. The visual acuity of the eye is about one minute of arc. If we optimistically assume half a minute, we can use a standard arc-length calculation to estimate the distance from which we would have to view the jump in order to resolve it. For a jump of 0.002 mm, this gives about 14 mm, or a little over a half-inch. The near point of the eye (the closest distance at which one can still focus) is about 25 cm, so we are out of luck.

References

Bracken, D.S., Rudy, C.R.: Principles and applications of calorimetric assay. Los Alamos National Laboratory report LA-UR-07-5226. http://www.lanl.gov/orgs/n/n1/panda/10.%20Calorimetry.pdf (2007)

DeGroot, G.: The Bomb: A Life. Harvard University Press, Cambridge, MA (2004). See p. 16

DOE: Historic American Engineering Record: B Reactor (105-B Building), HAER No. WA-164. http://www.cfo.doe.gov/me70/history/NPSweb/DOE-RL-2001-16.pdf (2001)

Glasstone, S., Dolan, P.J.: The Effects of Nuclear Weapons, 3rd edn. United States Department of Defense and Energy Research and Development Administration, Washington (1977)

Mark, J.C.: Explosive properties of reactor-grade plutonium. Sci. Glob. Secur. **4**(1), 111–128 (1993)

Reed, B.C.: Seeing the light: visibility of the July'45 Trinity Atomic Bomb Test from the inner solar system. Phys. Teach. **44**, 604–606 (2006)

Reed, B.C.: Can the energy of fission make a grain of sand visibly jump? Phys. Teach. **56**, 583 (2018)

Rhodes, R.: The Making of the Atomic Bomb. Simon and Schuster, New York (1986)

Serber, R.: The Los Alamos Primer: The First Lectures on How To Build An Atomic Bomb. University of California Press, Berkeley (1992)

Chapter 6
Appendices

6.1 Appendix A: Selected Δ-Values and Fission Barriers

Δ-values for nuclides involved in every reaction in this book are listed below. These are adopted from Jagdish K. Tuli, *Nuclear Wallet Cards* (Brookhaven National Laboratory, October 2011.) The full publication is available at www.nndc.bnl.gov. Fission barriers quoted for selected heavy nuclides are taken from P. Möller et al., "Fission barriers at the end of the chart of the nuclides," *Phys. Rev.* C **91**, 024310 (2015). These correspond to the listed nuclide as being the *compound* nucleus formed after neutron capture. For example, for fission of ^{235}U, the compound nucleus is ^{236}U, and the barrier is 5.03 MeV.

Nuclide	Δ (MeV)	Nuclide	Δ (MeV)	$E_{Barrier}$ (MeV)
$^{1}_{0}$n	8.071	$^{92}_{36}$Kr	−68.769	
$^{1}_{1}$H	7.289	$^{95}_{38}$Sr	−75.123	
$^{2}_{1}$H	13.136	$^{94}_{40}$Zr	−87.272	
$^{3}_{1}$H	14.950	$^{116}_{46}$Pd	−79.831	
$^{4}_{2}$He	2.425	$^{118}_{46}$Pd	−75.391	
$^{6}_{3}$Li	14.087	$^{139}_{54}$Xe	−75.644	
$^{7}_{3}$Li	14.907	$^{141}_{56}$Ba	−79.733	
$^{9}_{4}$Be	11.348	$^{150}_{66}$Dy	−69.310	
$^{10}_{5}$B	12.050	$^{206}_{82}$Pb	−23.786	
$^{12}_{6}$C	0.000	$^{208}_{82}$Pb	−21.749	
$^{13}_{6}$C	3.125	$^{210}_{84}$Po	−15.953	
$^{14}_{7}$N	2.863	$^{220}_{86}$Rn	10.607	
$^{16}_{8}$O	−4.737	$^{222}_{86}$Rn	16.373	
$^{17}_{8}$O	−0.809	$^{224}_{88}$Ra	18.821	
$^{17}_{9}$F	1.951	$^{226}_{88}$Ra	23.668	

(continued)

© The Author(s), under exclusive license to Springer Nature Switzerland AG 2021
B. C. Reed, *The Physics of the Manhattan Project*,
https://doi.org/10.1007/978-3-030-61373-0_6

(continued)

Nuclide	Δ (MeV)	Nuclide	Δ (MeV)	$E_{Barrier}$ (MeV)
$^{19}_{9}F$	−1.487	$^{231}_{91}Pa$	33.425	
$^{20}_{9}F$	−0.017	$^{232}_{91}Pa$	35.941	4.97
$^{20}_{10}Ne$	−7.042	$^{233}_{92}U$	36.921	4.79
$^{22}_{10}Ne$	−8.024	$^{235}_{92}U$	40.921	4.87
$^{26}_{10}Ne$	0.48	$^{236}_{92}U$	42.447	5.03
$^{25}_{12}Mg$	−13.192	$^{238}_{92}U$	47.310	5.63
$^{27}_{12}Mg$	−14.586	$^{239}_{92}U$	50.575	6.21
$^{27}_{13}Al$	−17.196	$^{237}_{93}Np$	44.874	4.94
$^{30}_{15}P$	−20.200	$^{239}_{93}Np$	49.313	5.57
$^{31}_{15}P$	−24.441	$^{239}_{94}Pu$	48.591	5.74
$^{35}_{16}S$	−28.846	$^{240}_{94}Pu$	50.128	5.98
$^{56}_{26}Fe$	−60.606	$^{241}_{95}Am$	52.937	6.34
		$^{242}_{95}Am$	55.471	6.72
		$^{252}_{99}Es$	77.29	6.79

6.2   Appendix B: Densities, Cross-Sections, Secondary Neutron Numbers, and Spontaneous-Fission Half-Lives

6.2.1   Thermal Neutrons (0.0253 eV)

Quantity	Unit	^{235}U	^{238}U	^{239}Pu	^{240}Pu
ρ	g cm^{-3}	18.71	18.95	15.6	15.6
A	g mol^{-1}	235.04	238.05	239.05	240.05
$\sigma_{capture}$	bn	98.81	2.717	270.3	289.4
$\sigma_{fission}$	bn	584.4	0	747.4	0.059
$\sigma_{elastic}$	bn	15.04	9.360	7.968	1.642
$\sigma_{inelastic}$	bn	0	0	0	0
ν	–	2.421	2.448	2.872	~3

6.2.2 Fast Neutrons (Fission-Spectrum Averages)

Quantity	Unit	^{235}U	^{238}U	^{239}Pu	^{240}Pu
$\sigma_{capture}$	bn	0.089	0.066	0.053	0.093
$\sigma_{fission}$	bn	1.235	0.308	1.800	1.357
$\sigma_{elastic}$	bn	4.566	4.804	4.394	4.319
$\sigma_{inelastic}$	bn	1.804	2.595	1.460	1.950
ν	–	2.637	2.655	3.172	2.257
$t_{1/2}$ (SF)	years	1.0×10^{19}	8.2×10^{15}	8.0×10^{15}	1.14×10^{11}

The density cited for ^{235}U, 18.71 g cm^{-3}, is (235/238) times that of natural uranium, 18.95 g cm^{-3}. Plutonium exhibits several different phases depending on temperature; the so-called "delta" phase is the one used for weapons (Bernstein 2007, 2008; Reed 2019, Chap. 7). The density figure for Pu cited here is that for the delta-phase as quoted on page 144 of Bernstein (2008). The ν value for ^{238}U for fission-energy neutrons is for neutrons of energy 2.9 MeV, and that for ^{240}Pu for fast neutrons is the number of neutrons emitted in the spontaneous fission of that isotope. Otherwise, secondary neutron numbers are adopted from ENDF files. Cross-sections are adopted from the Korean Atomic Energy Research Institute (KAERI) Table of Nuclides, http://atom.kaeri.re.kr/ton/index.htm. Cross-sections that are exceedingly small are recorded here as zero. For fast neutrons, cross-sections represent values averaged over the fission-energy spectrum.

6.3 Appendix C: Energy and Momentum Conservation in a Two-Body Collision

In many instances, we deal with reactions where an "incoming" nucleus strikes a second nucleus that is initially at rest, with two product nuclei emerging from the reaction. An example of this is the reaction used by Rutherford to induce the first-known artificial transmutation,

$$\tfrac{4}{2}\text{He} + \tfrac{14}{7}\text{N} \rightarrow \tfrac{1}{1}\text{H} + \tfrac{17}{8}\text{O}.$$

We are usually interested in the final kinetic energy and/or momentum of one of the product nuclei. In this section we develop formulae for these quantities, assuming that we are dealing with a head-on collision.

Figure 6.1 illustrates the situation. Let the *rest masses* of the nuclei be m_A, m_B, m_C, and m_D. Nucleus A is presumed to bring kinetic energy K_A into the reaction; the struck nucleus, B, is assumed to be at rest initially. Products C and D emerge from the reaction with kinetic energies K_C and K_D. If no transmutation is involved, we

Fig. 6.1 Head-on collision of two nuclei producing two other nuclei

can set $C = A$ and $D = B$. The goal here is to derive expressions for the final kinetic energies and momenta of nuclei C and D.

Begin with energy conservation. From the definition of Q in Sect. 1.1 we can write energy conservation for this reaction as

$$K_A = K_C + K_D - Q, \qquad (6.1)$$

where

$$Q = E_A + E_B - E_C - E_D, \qquad (6.2)$$

where the E's are the mc^2 rest energies of the nuclei.

As for momentum conservation, all reactions of this type that we will have occasion to examine will be non-relativistic. This allows us to deal with momentum from a purely classical perspective, which greatly simplifies the algebra. In Newtonian mechanics, the momentum p of a mass m which has kinetic energy K is given by $p = \sqrt{2mK} = \sqrt{2EK}\big/c$, so we have, upon canceling factors of 2 and c,

$$\sqrt{E_A K_A} = \pm\sqrt{E_C K_C} + \sqrt{E_D K_D}. \qquad (6.3)$$

A $\pm$ sign has been put in front of the momentum for nucleus C as a reminder that it may be moving forward or backward after the collision; the direction of C will emerge automatically from the analysis. *We assume that nucleus D is always moving forward after the reaction.*

The goal is to solve (6.1) and (6.3) for K_D in terms of the known quantities K_A, E_A, E_B, E_C, E_D, and Q. We need to eliminate K_C. To do this, first isolate the $\pm\sqrt{E_C K_C}$ term from (6.3) and square the result, which will cause the $\pm$ sign to disappear. Then solve (6.1) for K_C and substitute into the result of manipulating (6.3). The result is a quadratic in $\sqrt{K_D}$:

$$\alpha K_D + \beta \sqrt{K_D} + \gamma = 0, \qquad (6.4)$$

where

$$\alpha = (E_C + E_D), \qquad (6.5)$$

$$\beta = -2\sqrt{E_A E_D K_A}, \tag{6.6}$$

and

$$\gamma = (E_A K_A - E_C K_A - E_C Q). \tag{6.7}$$

Solving the quadratic gives

$$\sqrt{K_D} = \frac{-\beta \pm \sqrt{\beta^2 - 4\alpha\gamma}}{2\alpha}. \tag{6.8}$$

Provided that $\beta^2 - 4\alpha\gamma > 0$, there are two possible solutions for K_D, and either or both may be valid; see the next paragraph for further details on demanding a real solution of (6.8). The validities of the solutions can be checked *a posteriori* by computing K_C in two separate ways and checking for consistency: from (6.1), and from conservation of momentum by first computing $p_D = \sqrt{2m_D K_D} = \sqrt{2E_D K_D}/c$, demanding $p_C = p_A - p_D$, and then evaluating $K_C = \left(p_C^2/2E_C\right)c^2$.

In (6.8), a real solution will obtain for K_D only if $\beta^2 - 4\alpha\gamma > 0$. From (6.5)–(6.7), this demands

$$0 > K_A(E_A - E_C - E_D) - Q(E_C + E_D). \tag{6.9}$$

Now, $(E_A - E_C - E_D)$ is likely to be negative, so let us write (6.9) as

$$0 > -K_A|E_A - E_C - E_D| - Q(E_C + E_D). \tag{6.10}$$

Consider (6.10) in two separate cases: (i) $Q > 0$, and (ii) $Q < 0$. If $Q > 0$, we can write $Q = +|Q|$, and reduce (6.10) to

$$|Q|(E_C + E_D) > -K_A|E_A - E_C - E_D|. \tag{6.11}$$

You should be able to convince yourself that (6.11) is always true. This means that in cases where $Q > 0$, there is no constraint on K_A. On the other hand, if $Q < 0$, write $Q = -|Q|$, in which case (6.10) gives

$$0 > -K_A|E_A - E_C - E_D| + |Q|(E_C + E_D), \tag{6.12}$$

which demands

$$K_A > \frac{|Q|(E_C + E_D)}{|E_A - E_C - E_D|} \quad (Q < 0) \tag{6.13}$$

This expression means that there is a *threshold* energy for K_A in cases where $Q < 0$.

Table 6.1 Rutherford alpha-bombardment reaction parameters

Reactant	Nuclide	A	Δ (MeV)	Rest mass (MeV/c^2)
A	$^4_2$He	4	2.425	3728.401
B	$^{14}_7$N	14	2.863	13043.779
C	$^{17}_8$O	17	−0.809	15834.589
D	$^1_1$H	1	7.289	938.783

We now apply this analysis to Rutherford's transmutation reaction. Identify A, B, C, and D as He, N, O, and H, respectively. The relevant numbers appear in Table 6.1. This reaction has a Q-value of −1.192 MeV. The conversion factor $\varepsilon = 931.494$ MeV amu^{-1} was used to compute rest masses in MeV/c^2 via the relationship (rest mass) $= \varepsilon A + \Delta$.

Suppose that the alpha particle enters the reaction with $K_A = 5$ MeV. Then we have

$$\alpha = (E_C + E_D) = 16773.37 \quad \text{(MeV)},$$

$$\beta = -2\sqrt{E_A E_D K_A} = -8366.79 \quad \text{(MeV)}^{3/2},$$

and

$$\gamma = (E_A K_A - E_C K_A - E_C Q) = -41656.11 \quad \text{(MeV)}^2$$

The two solutions for K_D give 3.404 and 1.812 MeV. The first of these proves to be physically valid, but the second does not because it fails the consistency check for K_C. The corresponding momentum of the proton is

$$p_D = \frac{\sqrt{2E_D K_D}}{c} = \frac{1}{c}\sqrt{2\,(938.783 \text{ MeV})\,(3.404 \text{ MeV})} = 79.94\frac{\text{MeV}}{c}.$$

The oxygen nucleus emerges from the reaction with kinetic energy $K_C = 0.404$ MeV and momentum 113.15 MeV/c.

Equation (6.13) gives a threshold energy of $K_A > 1.533$ MeV for this reaction. This value is greater than 1.192 MeV because *both* momentum and energy must be conserved; were nucleus A to strike nucleus B with only 1.192 MeV of kinetic energy, nuclei C and D would emerge from the reaction with no kinetic energy and hence no momentum, a situation inconsistent with A bringing momentum into the reaction in the first place. These calculations are carried out in the spreadsheet **TwoBody.xls**.

6.4 Appendix D: Energy and Momentum Conservation in a Two-Body Collision that Produces a Gamma-Ray

In Sect. 1.4, we examined the Joliot-Curies' proposed gamma-producing reaction

$$\ce{^4_2He + ^9_4Be -> ^{13}_6C} + \gamma \tag{6.14}$$

The alpha-particle carries ~5.3 MeV of kinetic energy into the reaction, and collides with the initially stationary Be nucleus. The quantities of interest in this reaction are the energy and momentum of the emergent gamma-ray. In this section we develop formulae for these quantities, assuming that the collision is head on.

Figure 6.2 illustrates the situation. Let the *rest masses* of the three nuclei be m_A, m_B, and m_C. Nucleus A is presumed to bring kinetic energy K_A into the reaction. Product C emerges from the reaction with kinetic energy K_C and the gamma-ray with energy E_γ. We do not refer to E_γ as a kinetic energy as that term is usually reserved for the motional energy of a particle of non-zero rest mass.

Begin with energy conservation, accounting for the kinetic energies of the reactants as well as their relativistic rest energies:

$$K_A + (m_A + m_B)\,c^2 = K_C + m_C c^2 + E_\gamma. \tag{6.15}$$

For momentum conservation. we take the Newtonian momentum $p = \sqrt{2mK}$ for particles with mass, and, from Einstein, $p = E/c$ for the gamma-ray, giving

$$\sqrt{2m_A K_A} = \pm\sqrt{2m_C K_C} + E_\gamma\big/c, \tag{6.16}$$

where the upper (lower) sign is to be taken if C is moving to the right (left) after the reaction; we assume that the gamma-ray is always moving forward after the reaction. The direction of C after the reaction is dictated by energy and momentum conservation. Let the mc^2 rest energies of the particles with mass be designated by E's, e.g., $E_A = m_A c^2$. If we replace the masses in (6.16) by these rest energies, a factor of c can be canceled, leaving

$$\sqrt{2E_A K_A} = \pm\sqrt{2E_C K_C} + E_\gamma. \tag{6.17}$$

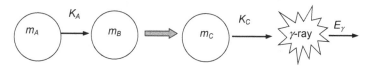

Fig. 6.2 Head-on collision of two massive particles leading to production of a massive particle and a gamma-ray

We desire to eliminate K_C between (6.15) and (6.17). Rearrange (6.17) to isolate the $\sqrt{2E_C K_C}$ term, square, and then solve for K_C. The $\pm$ sign vanishes, and we get

$$K_C = \left(\frac{E_A}{E_C}\right) K_A - \frac{\sqrt{2E_A K_A}\, E_\gamma}{E_C} + \frac{E_\gamma^2}{2E_C}. \tag{6.18}$$

Now rearrange (6.15) to the form

$$K_C = E_A + E_B + K_A - E_C - E_\gamma. \tag{6.19}$$

Substitute (6.19) into (6.18) and rearrange; the result is a quadratic equation in E_γ:

$$\alpha\, E_\gamma^2 + \varepsilon\, E_\gamma + \delta = 0, \tag{6.20}$$

where

$$\alpha = \frac{1}{2E_C}, \tag{6.21}$$

$$\varepsilon = 1 - \frac{\sqrt{2E_A K_A}}{E_C}, \tag{6.22}$$

and

$$\delta = \left(\frac{E_A}{E_C}\right) K_A - (E_A + E_B + K_A - E_C). \tag{6.23}$$

Hence,

$$E_\gamma = \frac{-\varepsilon \pm \sqrt{\varepsilon^2 - 4\alpha\delta}}{2\alpha}. \tag{6.24}$$

There are two possible solutions for E_γ; often, one will be unphysical in that it leads to a negative value for K_C from (6.19). From (6.21)–(6.23), we can expect in any sensible circumstance to find that α will be small, whereas ε will have a value close to unity (unless A comes into the reaction with a relativistic amount of kinetic energy), and δ will be less than zero, a combination which will require always taking the positive root in (6.24). These calculations are done in the spreadsheet **TwoBodyGamma.xls**.

6.5 Appendix E: Formal Derivation of the Bohr-Wheeler Spontaneous Fission Limit

6.5.1 Introduction

In Sects. 1.7 and 1.10, we used a simplified model of a fissioning nucleus to get a sense of how the limit against spontaneous fission (SF), $(Z^2/A) = 2(a_S/a_C) \sim 48$, arises, a result first derived by Bohr and Wheeler (1939). Given the significance of this result, a formal derivation of it is presented here. This approach is somewhat unusual in comparison to most texts, which do not present detailed derivations of this work. Some do offer partial treatments based on starting from expressions for the surface area and self-energy of an ellipsoid of variable eccentricity (see, for example, Bernstein and Pollock 1979 or Cottingham and Greenwood 2001), but the ellipsoidal model does not reflect the approach taken by B&W, who used a sum of Legendre polynomials to describe the shape of the surface of a distorted nucleus. While it is true that it should not matter how the distortion is modeled if the SF limit is a matter of instability against slight distortions, it seems unfortunate that pedagogical tendency has shifted away from historical accuracy.

The popularity of the ellipsoidal model is due to the fact that the mathematics of the B&W analysis *is* tricky, even if one is facile with multivariable calculus and Legendre polynomials. B&W published virtually none of the details of their work, which they referred to as a "straightforward calculation." Soon after B&W's paper appeared, Present and Knipp (1940a, b) pointed out that it contained an internal inconsistency, and that B&W had changed the definition of some of the surface-distortion parameters part-way through the derivation. In a paper that now seems largely forgotten, Plesset (1941) reconstructed the details of the B&W derivation, but his work is difficult to follow in view of some tangled notation and the fact that he carried through his algebra to higher orders of perturbation than are necessary to understand the SF limit.

In reconstructing the B&W derivation, one faces the question of what level of detail to present. To lay out every step of the algebra would be far too lengthy for sensible publication. Conversely, the danger of brevity is that subtle but important points can get overlooked. Here I try to tread a middle path by setting down benchmark steps in the calculations between which readers should be able fill in the intervening details. No treatment is given here of the much more complex question of the fission *barrier*, which requires carrying the algebra to higher orders of perturbation.

This derivation, which is adopted from Reed (2009), is rather lengthy. In Sect. 6.5.2, the Legendre-polynomial model of a distorted nucleus is described, and the calculation of the volume of the nucleus is carried out. The surface area energy is calculated in Sect. 6.5.3. Section 6.5.4 deals with the calculation of the Coulomb self-energy of the nucleus, which, when combined with the results of the preceding sub-sections, leads to understanding how the SF limit arises.

6.5.2 *Nuclear Surface Profile and Volume*

Bohr and Wheeler began by imagining an initially spherical nucleus of radius R_O undergoing a distortion expressible in the form

$$r(\theta) = R_O\{(1 + \alpha_0)P_0(\cos\theta) + \alpha_2 P_2(\cos\theta) + \cdots\}. \qquad (6.25)$$

$r(\theta)$ is the shape of the nucleus as a function of the spherical polar angle θ; see the sketch in Fig. 6.3. $P_0(\cos\theta)$ and $P_2(\cos\theta)$ are respectively zeroth- and second-order Legendre polynomials.

Most students become familiar with Legendre polynomials in the study of electromagnetism or quantum mechanics. These polynomials are an infinite family of functions of an argument which in most physical applications is the cosine of the spherical polar angle θ. The subscript on the P designates the highest order of the argument which appears in the polynomial. Our attention will be restricted to the first three such polynomials,

$$P_0(\cos\theta) = 1, \qquad (6.26)$$

$$P_1(\cos\theta) = \cos\theta, \qquad (6.27)$$

Fig. 6.3 The surface of a distorted nucleus is described by the function $r(\theta)$ of (6.25). A ribbon of surface of edge length ds and area $dS = 2\pi\, r\, sin\theta\, ds$ at colatitude θ is shown

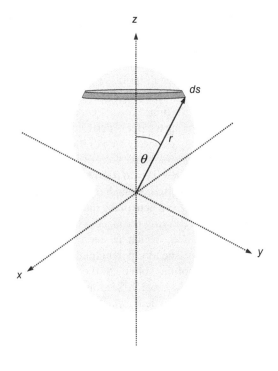

and

$$P_2(\cos\theta) = \frac{1}{2}(3\cos^2\theta - 1).\tag{6.28}$$

Such a perturbation as (6.25), greatly exaggerated, is sketched schematically in Fig. 6.3, where the nucleus has been perturbed into a dumbbell shape along the polar axis; see also Fig. 6.4. The coefficients α_0 and α_2 are presumed to be small; using only two coefficients is sufficient to derive the SF limit. Coefficient α_2 dictates the non-spherical shape of the nucleus; α_0 is necessary to be able to ensure volume conservation as the distortion occurs. It is conventional to consider α_2 as the "independent" coefficient, and ultimately express both the area and Coulomb energies as functions of it alone.

What might cause a nucleus to become distorted in the first place? In the case of a uranium nucleus struck by a neutron, the collision itself will presumably introduce some distortion; if the binding energy released exceeds the fission barrier, then the nucleus will proceed to fission. But what about a nucleus that is just sitting around minding its own business? Here it is necessary to appreciate that nuclei are not the hard, static, billiard-ball-like spheres of elementary-school imagination; protons and neutrons exert tremendous forces on each another and so nuclei are in constant states of roiling agitation. If a group of protons and neutrons should find themselves temporarily forming an alpha-particle or even larger sub-nucleus, quantum tunneling can cause alpha-decay or spontaneous fission to occur. Also, no nucleus can ever be removed from all outside influences: Even in the depths of interstellar space, they will be under constant bombardment from background photons.

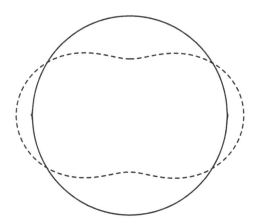

Fig. 6.4 The solid line shows (in cross-section) a spherical nucleus with $R_O = 1$; the dashed line shows the same nucleus distorted to $\alpha_2 = $ -0.4 and $\alpha_0 = -\alpha_2^2/5 = $ -0.032 (see Eq. 6.36). This is by no means a "small" perturbation, but is deliberately chosen to show an appreciable distortion. The extent of the distorted nucleus is to $r \sim \pm 1.16$ across the equator; at the poles the distortion goes to $r \sim \pm 0.57$

The essence of the Bohr-Wheeler calculation is to compare the total energy of the deformed nucleus ($\alpha_2 \neq 0$) to that which it had in its initial spherical condition ($\alpha_0 = \alpha_2 = 0$), and then to determine what circumstance must hold so that *any* perturbation, no matter how slight, will yield a lower-energy configuration toward which the nucleus would presumably proceed spontaneously. The lowest-order contributions to these energies both prove to be of order α_2^2; it is not necessary to carry through the algebra to any higher orders to establish the SF limit. Some texts do not emphasize that the volume of the nucleus is assumed to be conserved, that is, that nuclei are considered to be incompressible.

Note that there is no "first-order" term $\alpha_1 P_1 = \alpha_1 \cos\theta$ in (6.25). The reason for this is sometimes stated as being that such a term (or, indeed, any odd-parity perturbation) creates only a displacement of the center of mass of the nucleus along the z-axis, but this is not quite the whole story. Such a term would introduce a distortion of the shape of the nucleus, rendering it somewhat flattened at the "south pole" ($\theta = \pi$) if $\alpha_1 > 0$. Incorporating only even-order perturbations from sphericity simplifies the situation to having a nucleus whose center of mass remains at the coordinate origin, and which remains symmetric about the xy plane. Because $r(\theta)$ contains no dependence on the azimuthal angle ϕ, the nucleus also remains symmetric about the polar axis. The sign of α_2 dictates the nature of the distortion. If $\alpha_2 > 0$, the nucleus becomes squeezed at the equator and elongated at the poles, as suggested in Fig. 6.3; $\alpha_2 < 0$ produces the opposite effect, rendering the nucleus somewhat doughnut-shaped in the equatorial plane (Fig. 6.4). The first term in (6.25) could be written as $(1 + \alpha_0)$ since P_0 (cos θ) $= 1$, but I will write out the P's for sake of explicitness.

Why use Legendre polynomials? The surface of the nucleus could presumably be described by any arbitrarily-chosen function of the spherical coordinates (θ, ϕ), subject only to the condition that the function contains enough parameters to accommodate ensuring conservation of volume. The value of Legendre polynomials is that, as described below, integrals of products of them satisfy certain orthogonality relationships. These relationships greatly simplify calculations of quantities such as the surface area and volume of such a distorted shape. In any context where one needs to model perturbations from circularity or sphericity, Legendre polynomials are a convenient family of functions for doing so.

The first task is to ensure conservation of volume. The volume of the distorted nucleus is given by

$$V = \int\limits_{\theta=0}^{\pi} \int\limits_{r=0}^{r(\theta)} \int\limits_{\phi=0}^{2\pi} r^2 \sin\theta \, d\phi \, dr \, d\theta. \tag{6.29}$$

Note carefully the order of integrations over r and θ. Because the upper limit of r is a function of θ, the integral over r must be done first, then that over θ. The integral over ϕ gives 2π directly; this will be the case for any integral over ϕ in what follows. Hence we have

$$V = \frac{2\pi}{3} \int\limits_{\theta=0}^{\pi} r^3(\theta) \sin\theta \, d\theta. \tag{6.30}$$

Be sure to understand the distinction between the integrands in (6.29) and (6.30). In (6.29), r is a variable whose limits are 0 and $r\,(\theta)$; the $r^3\,(\theta)$ in (6.30) means r as a function of θ as given by the cube of (6.25).

The B&W calculation involves numerous integrals of the form of (6.30), with various powers of $r(\theta)$ and often other functions of θ in the integrand. To simplify notation, it is convenient to make a change of variable to $x = cos\theta$, which renders $sin\theta\ d\theta$ as $-dx$, with limits $x = (1,-1)$. (Note that x is not the usual Cartesian coordinate, but rather just a transformation variable.) The limits can be flipped, with the result that the negative sign in $-dx$ can be dropped. In terms of this formulation, integrals of products of two Legendre polynomials work out very simply. In general, if the P's are of different orders, then the integral of their product over $x = (-1, 1)$ is identically zero:

$$\int_{-1}^{1} P_i\, P_j\, dx = 0, \qquad (i \neq j). \tag{6.31}$$

If the integral involves the product of a given P with itself, the result is

$$\int_{-1}^{1} P_n^2\, dx = \frac{2}{2n + 1}. \tag{6.32}$$

As a check, you might wish to verify that various combinations of the specific cases of (6.26)–(6.28) satisfy (6.31) and (6.32). Note that (6.31) and (6.32) do *not* apply if there are other functions of θ in the integrands in addition to the Legendre polynomials.

We can now evaluate the volume integral (6.30). Transforming to x and substituting (6.25) into (6.30) gives

$$V = \left(\frac{2\pi\, R_O^3}{3}\right) \int_{-1}^{1} [(1 + \alpha_0)P_0 + \alpha_2 P_2]^3\, dx, \tag{6.33}$$

where we suppress the $(cos\,\theta)$ arguments of the P's for brevity. Treating α_0 and α_2 as constants and cubing gives

$$V = \left(\frac{2\pi\, R_O^3}{3}\right) \left\{ (1 + \alpha_0)^3 \int_{-1}^{1} P_0^3 dx + 3\,(1 + \alpha_0)^2 \alpha_2 \int_{-1}^{1} P_0^2 P_2\, dx \right.$$
$$\left. + 3\,(1 + \alpha_0)\alpha_2^2 \int_{-1}^{1} P_0 P_2^2 dx + \alpha_2^3 \int_{-1}^{1} P_2^3 dx \right\}. \tag{6.34}$$

By (6.32), the first here integral gives 2 since you can imagine extracting one factor of $P_0 = 1$ out in front of the integral to leave two such factors inside. This is an important point: *Since $P_0 = 1$, we can always extract a factor of P_0 from within an integral, or, equivalently, multiply any integrand we come across by P_0 as is convenient; this can be a handy trick to see if an integral will vanish by virtue of* (6.31). The second integral in (6.34) vanishes via precisely this trick, and the third integral gives 2/5 by (6.32). The last integral gets dropped as we retain terms only to order α_2^2. To this order, the volume evaluates as

$$V = \left(\frac{4\pi R_O^3}{3}\right)\left\{(1+\alpha_0)^3 + \frac{3}{5}(1+\alpha_0)\alpha_2^2 + \cdots\right\}. \tag{6.35}$$

If volume is to be conserved, then the contents of the brace bracket in (6.35) must equal unity. If α_0 and α_2 are presumed small, then the $\alpha_0\alpha_2^2$ and α_0^3 terms can be dropped; what remains is a quadratic equation in α_0 whose solution is

$$\alpha_0 \sim -\frac{1}{5}\alpha_2^2. \tag{6.36}$$

This result will prove valuable in computing the area and Coulomb energies.

6.5.3   The Area Integral

Figure 6.3 shows a ribbon of surface area at spherical-coordinate polar angle θ and angular width $d\theta$ that goes all the way around the nucleus. The area of the ribbon will be its arc length times its circumference $2\pi r(\sin\theta)$. However, the deformed nucleus does not have a spherical profile, so the arc length is not simply $r\,d\theta$. Rather, we have to compute it by using the general expression for arc-length in spherical coordinates for a trajectory running along a line of constant "longitude" ϕ:

$$ds^2 = dr^2 + r^2 d\theta^2. \tag{6.37}$$

Since r is a function of θ, we can write this as

$$ds^2 = dr^2 + r^2 d\theta^2 = r^2 d\theta^2\left[1 + \frac{1}{r^2}\left(\frac{dr}{d\theta}\right)^2\right]. \tag{6.38}$$

The area of the ribbon dS is then

$$dS = 2\pi\, r\, \sin\theta\, ds = 2\pi\, r^2\, \sin\theta\, \sqrt{1 + \frac{1}{r^2}\left(\frac{dr}{d\theta}\right)^2}\, d\theta. \tag{6.39}$$

If the nucleus is not greatly distorted, then $dr/d\theta$ will be small. We can then invoke a binomial expansion,

$$\sqrt{1 + \frac{1}{r^2}\left(\frac{dr}{d\theta}\right)^2} \sim 1 + \frac{1}{2}\frac{1}{r^2}\left(\frac{dr}{d\theta}\right)^2 - \frac{1}{8}\frac{1}{r^4}\left(\frac{dr}{d\theta}\right)^4 + \cdots \tag{6.40}$$

From (6.25), $(dr/d\theta) = R_O\, \alpha_2\, (d P_2/d\theta)$, so, to retain terms to order α_2^2, we need only carry the first two terms in the expansion in (6.40):

$$dS = 2\pi\, \sin\theta\, \left\{ r^2 + \frac{1}{2}\left(\frac{dr}{d\theta}\right)^2 + \cdots \right\} d\theta. \tag{6.41}$$

To this level of approximation, the surface area of the deformed nucleus comprises two contributions:

$$S = 2\pi\left\{ \int_0^\pi r^2 \sin\theta\, d\theta + \frac{1}{2}\int_0^\pi \left(\frac{dr}{d\theta}\right)^2 \sin\theta\, d\theta + \cdots \right\}. \tag{6.42}$$

Using (6.31) and (6.32), these integrals reduce to

$$S \sim 4\pi\, R_O^2 \left\{ (1 + \alpha_0)^2 + \frac{4}{5}\alpha_2^2 + \cdots \right\}. \tag{6.43}$$

Substitute into this the result of volume conservation, $\alpha_0 \sim -\alpha_2^2/5$. Also invoke the usual nuclear radius approximation $R_O \sim a_0 A^{1/3}$ ($a_0 \sim 1.2$ fm), and write the factor which converts surface area to equivalent energy as Ω. The surface energy U_S can then be written as

$$U_S \sim \left(a_S A^{2/3}\right)\left\{ 1 + \frac{2}{5}\alpha_2^2 + \cdots \right\}, \tag{6.44}$$

where $a_S = 4\pi\,\Omega\, a_0^2 \sim 18$ MeV. The areal energy *increases* upon perturbation of the nucleus from its initially spherical shape; this is understandable in that a sphere is the surface of minimum area which encloses a given volume.

6.5.4 The Coulomb Integral and the SF Limit

Figure 6.5 illustrates the geometry of computing the Coulombic self-potential of the distorted nucleus.

The nucleus is divided into elements of volume $d\tau$; protons are assumed to be uniformly distributed throughout the nucleus, leading to a constant charge density ρ. By considering pairs of volume elements labeled as "1" and "2", the electrostatic self-energy is computed from

$$U_C = \frac{1}{2}\left(\frac{\rho^2}{4\pi\varepsilon_0}\right) \int\limits_{(1)} \int\limits_{(2)} \frac{d\tau_1 \, d\tau_2}{r_{12}}, \qquad (6.45)$$

where r_{12} is the distance between the two volume elements. Each volume element is three-dimensional, so (6.45) is actually a sextuple integral. As in the computation of the surface area, integrals over r must be done before those over θ. Care must be taken to keep track of "1" and "2" integrals and coordinates.

To treat the factor of r_{12} in the denominator of (6.45), apply the law of cosines to the triangle r_1-r_2-r_{12} in Fig. 6.5, and carry out a binomial expansion for two separate cases: $r_2 < r_1$, and $r_2 > r_1$. This gives

Fig. 6.5 Geometry for computing the Coulomb self-energy of the distorted nucleus. The two volume elements are located at distances r_1 and r_2 from the origin, and are separated by distance r_{12}. The angle between them as viewed from the origin is θ_{12}

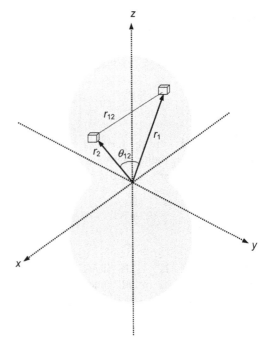

$$\frac{1}{r_{12}} = \begin{cases} \dfrac{P_0(\cos\theta_{12})}{r_1} + \left(\dfrac{r_2}{r_1^2}\right)P_1(\cos\theta_{12}) + \left(\dfrac{r_2^2}{r_1^3}\right)P_2(\cos\theta_{12}) + \cdots & (r_2 < r_1) \\\\ \dfrac{P_0(\cos\theta_{12})}{r_2} + \left(\dfrac{r_1}{r_2^2}\right)P_1(\cos\theta_{12}) + \left(\dfrac{r_1^2}{r_2^3}\right)P_2(\cos\theta_{12}) + \cdots & (r_2 > r_1), \end{cases}$$

$$(6.46)$$

where the ellipses indicate terms of fourth power and higher in the factors of r_1 and r_2 in the denominators. Note carefully that θ_{12} is the angle *between* the directions from the origin to volume elements 1 and 2, not either of the individual orientation angles θ_1 or θ_2. This is an important point: The Legendre polynomials are functions of a generic argument $\cos\theta$; as long as the pattern of factors of the argument appear as in (6.26)–(6.28), one has Legendre polynomials. These are exactly the patterns θ_{12} that turn up upon performing the above binomial expansion. The patterns in (6.46) persist into higher-order Legendre polynomials, but keeping just the terms written out here will be enough for our purposes.

It is immaterial whether one integrates over the "1" or "2" coordinates first. I elect the latter, and proceed by writing (6.45) as

$$U_C = \frac{\rho^2}{8\pi\varepsilon_0} \int\limits_{(1)} \left\{ \int\limits_{(2)} \frac{d\tau_2}{r_{12}} \right\} d\tau_1. \tag{6.47}$$

Call the inner integral U_2. To proceed, break it into two regimes, one for $0 \le r_2 \le r_1$ (for which $r_2 < r_1$, always), and another for which $r_1 \le r_2 \le r_2(\theta_2)$:

$$U_2 = \int\limits_{(2)} \frac{d\tau_2}{r_{12}} = \underbrace{\int\limits_{\theta,\,\phi} \int\limits_0^{r_1} \frac{d\tau_2}{r_{12}}}_{U_{21}(r_2 < r_1)} + \underbrace{\int\limits_{\theta,\,\phi} \int\limits_{r_1}^{r_2(\theta_2)} \frac{d\tau_2}{r_{12}}}_{U_{22}(r_2 > r_1)}, \tag{6.48}$$

where $d\tau_2 = r_2^2 dr_2\, d\Omega_2 = r_2^2 dr_2\, \sin\theta_2\, d\theta_2\, d\phi_2$ is the volume element in "2" coordinates. For $1/r_{12}$, use (6.46) as appropriate. To keep the algebra tractable, we will do the two integrals separately, referring to them as U_{21} and U_{22} as indicated. For U_{21} we have

$$U_{21} = \int\limits_{\theta,\,\phi} \int\limits_0^{r_1} \left\{ \frac{P_0(\theta_{12})}{r_1} + \left(\frac{r_2}{r_1^2}\right)P_1(\theta_{12}) + \left(\frac{r_2^2}{r_1^3}\right)P_2(\theta_{12}) \right\} r_2^2\, dr_2\, d\Omega_2. \tag{6.49}$$

In these integrals, θ_{12} is a shorthand for $\cos(\theta_{12})$. Since these integrals are over the "2" coordinates, factors of r_1 can be extracted from within them; we will eventually write $r_1(\theta_1)$. For the integration over r_2 we then get simple powers of r_2, which become powers of r_1 upon substituting the limits:

$$U_{21} = r_1^2 \int\limits_{\phi=0}^{2\pi} \int\limits_{\theta=0}^{\pi} \left\{ \frac{1}{3} P_0(\theta_{12}) + \frac{1}{4} P_1(\theta_{12}) + \frac{1}{5} P_2(\theta_{12}) \right\} d\Omega_2. \tag{6.50}$$

We come now to a very important step in the algebra. This is that there exists an identity known as the Addition Theorem for spherical harmonics. This theorem lets one write a Legendre polynomial $P_k(\cos \theta_{12})$ in terms of products of so-called Associated Legendre polynomials $P_k^m(\cos \theta)$ whose arguments are the cosines of the *individual* direction angles of the volume elements. If $m = 0$, the Associated Legendre polynomials become the "ordinary" Legendre polynomials of (6.26)–(6.28); this will come to be the case in a moment. The Addition Theorem is

$$P_k(\cos \theta_{12}) = \sum_{m=-k}^{k} \frac{(k-m)!}{(k+m)!} P_k^m(\cos \theta_1) P_k^m(\cos \theta_2) \exp[\imath m(\phi_1 - \phi_2)]. \tag{6.51}$$

This paragraph is important; read it carefully. Imagine (6.51) substituted into (6.50) where the various P's appear; also remember that we must eventually integrate over coordinate set "1" in (6.47). When integrating over ϕ_1 and ϕ_2, only $m = 0$ will give non-zero contributions because of the imaginary exponential in (6.51); you should convince yourself of the veracity of this statement. The Associated Legendre polynomials consequently reduce to regular Legendre polynomials, which I will designate as $P_{k(1)}$ and $P_{k(2)}$, where $P_{k(1)}$ designates the k'th-order Legendre polynomial for coordinate set "1", and $P_{k(2)}$ that for coordinate set "2".

The value of invoking the Addition Theorem is that we reduce U_{21} in (6.50) to products of individual-angle P's, which allows us to invoke (6.31) and (6.32). Now set $d\Omega_2 = \sin\theta_2 \, d\theta_2 \, d\phi_2$ in (6.50); do not forget the factor of 2π from integrating over ϕ. This gives

$$U_{21} = 2\pi r_1^2 \left\{ \frac{P_{0(1)}}{3} \int\limits_{\theta=0}^{\pi} P_{0(2)} \sin\theta_2 d\theta_2 + \frac{P_{1(1)}}{4} \int\limits_{\theta=0}^{\pi} P_{1(2)} \sin\theta_2 d\theta_2 \right.$$

$$\left. + \frac{P_{2(1)}}{5} \int\limits_{\theta=0}^{\pi} P_{2(2)} \sin\theta_2 d\theta_2 \right\}. \tag{6.52}$$

Invoking (6.32) indicates that the last two integrals in (6.52) vanish; recall that you can insert a factor of $P_{0(2)} = 1$ into any integrand as desired. The first integral in (6.52) is equal to 2 via (6.31), hence

$$U_{21} = \frac{4\pi r_1^2}{3} P_{0(1)}.$$

(6.53)

Integral U_{22} in (6.48) proceeds similarly, but with one important exception: The upper limit of $r_2(\theta_2)$ means that we must substitute (6.25) before integrating over θ_2. Carrying out the integral over r_2 and again invoking the Addition Theorem with restriction to $m = 0$ gives

$$U_{22} = \underbrace{\frac{P_{0(1)}}{2} \int_{\theta,\phi} \left\{ r_2^2(\theta_2) - r_1^2 \right\} P_{0(2)} d\Omega_2}_{U_{22A}} + \underbrace{r_1 P_{1(1)} \int_{\theta,\phi} \left[r_2(\theta_2) - r_1 \right] P_{1(2)} d\Omega_2}_{U_{22B}}$$

$$+ \underbrace{r_1^2 P_{2(1)} \int_{\theta,\phi} \left\{ \ln[r_2(\theta_2)] - \ln(r_1) \right\} P_{2(2)} d\Omega_2}_{U_{22C}}.$$

(6.54)

In all of these sub-integrals, r_1 can be regarded as a constant since we are integrating over coordinate set "2".

Integral U_{22B} proves to vanish: Substitute (6.25) for $r_2(\theta_2)$, writing the P's as $P_{0(2)}$ and $P_{2(2)}$; remember that it is legal to multiply the r_1 term by $P_{0(2)} = 1$. With the factor of $P_{1(2)}$ in the integrand, the products of the various P's are all guaranteed to integrate to zero by (6.31).

For integral U_{22A}, the r_1^2 term immediately integrates to $-4\pi r_1^2$ since we can imagine that r_1^2 is multiplied by $P_{0(2)} = 1$. For the $r_2^2(\theta_2)$ term, first square (6.25) and then carry out the resulting integrals using (6.31) and (6.32). You will find that a $P_{0(2)}^3$ term arises, but this can be dealt with by extracting one factor of $P_{0(2)}$ to the front of the integral as was done when computing the volume of the distorted nucleus. The result is

$$U_{22A} = 2\pi R_O^2 \, P_{0(1)} \left[(1 + \alpha_0)^2 + \frac{1}{5}\alpha_2^2 \right] - 2\pi r_1^2 P_{0(1)}.$$

(6.55)

In integral U_{22C}, the term involving $\ln(r_1)$ evaluates to zero because r_1 acts as a constant for an integral over "2" coordinates and we can multiply it by $P_{0(2)} = 1$; this leads to the product $P_{0(2)} P_{2(2)}$ and hence a zero result by (6.32).

The $\ln[r_2(\theta_2)]$ term in U_{22C} is trickier to evaluate. Begin by writing out $\ln[r_2(\theta_2)]$ using (6.25). Then extract a factor of $(1 + \alpha_0)$ from within the logarithm, and use the fact that the logarithm of a product is the sum of the logarithms of the terms in the product:

$$U_{22C} = r_1^2 P_{2(1)} \left\{ \ln[R_O(1+\alpha_0)] \int\limits_{\theta,\phi} P_{2(2)} d\Omega_2 \right.$$

$$\left. + \int\limits_{\theta,\phi} \ln\left[1 + \frac{\alpha_2 P_{2(2)}}{(1+\alpha_0)}\right] P_{2(2)} d\Omega_2 \right\}. \tag{6.56}$$

The first integral in (6.56) vanishes by (6.32) because we can insert a factor of $P_{0(2)}$ in the integrand. For the second integral, if α_2 is small, then we have an integrand of the form $\ln(1 + x)$ where x will be a small quantity. To deal with this, invoke the expansion

$$\ln(1 + x) \quad \sim \quad x - \frac{1}{2}x^2 + \cdots \tag{6.57}$$

This gives

$$U_{22C} = r_1^2 P_{2(1)} \left\{ \frac{\alpha_2}{(1+\alpha_0)} \int\limits_{\theta,\phi} P_{2(2)} P_{2(2)} d\Omega_2 \right.$$

$$\left. - \frac{\alpha_2^2}{2(1+\alpha_0)^2} \int\limits_{\theta,\phi} P_{2(2)}^3 d\Omega_2 \right\}. \tag{6.58}$$

The first integral in (6.58) evaluates to $4\pi/5$. The second one involves the cube of $P_{2(2)}$. It turns out that we will not need this second term, but we will carry it along for the time being with a compacting notation and write U_{2CC} in the form

$$U_{22C} = r_1^2 P_{2(1)} \left\{ \frac{4\pi}{5}\left(\frac{\alpha_2}{1+\alpha_0}\right) - \frac{\pi\alpha_2^2}{(1+\alpha_0)^2}[2,2,2] \right\}, \tag{6.59}$$

where

$$[2,2,2] = \int\limits_{\theta,\phi} P_{2(2)}^3 d\Omega_2. \tag{6.60}$$

Gathering (6.53), (6.55), and (6.59) into (6.48) gives

$$U_2 = -\frac{2\pi}{3}r_1^2 P_{0(1)} + 2\pi R_O^2 \, P_{0(1)}\left[(1+\alpha_0)^2 + \frac{1}{5}\alpha_2^2\right]$$

$$+ r_1^2 P_{2(1)} \left\{ \frac{4\pi}{5}\left(\frac{\alpha_2}{1+\alpha_0}\right) - \frac{\pi\alpha_2^2}{(1+\alpha_0)^2}[2,2,2] \right\}. \tag{6.61}$$

At this point, (6.61) goes back into (6.47) to give the overall Coulomb self-energy of the nucleus as

$$
U_C = \frac{\rho^2}{8\pi\varepsilon_o} \left\{ -\frac{2\pi}{3} \int\limits_{\theta,\phi} \int\limits_0^{r_1(\theta)} P_{0(1)} r_1^4 dr_1 d\Omega_1 + \right.
$$

$$
+ 2\pi R_O^2 \left[(1+\alpha_0)^2 + \frac{1}{5}\alpha_2^2 \right] \int\limits_{\theta,\phi} \int\limits_0^{r_1(\theta)} P_{0(1)} r_1^2 dr_1 d\Omega_1
$$

$$
\left. + \left[\frac{4\pi}{5}\left(\frac{\alpha_2}{1+\alpha_0}\right) - \frac{\pi\alpha_2^2}{(1+\alpha_0)^2}[2,2,2] \right] \int\limits_{\theta,\phi} \int\limits_0^{r_1(\theta)} P_{2(1)} r_1^4 dr_1 d\Omega_1 \right\}. \quad (6.62)
$$

The integrals in (6.62) proceed as did the U_2 integrals above. Integrating over r_1 in the first integral will yield $r_1^5(\theta)/5$, but it is not necessary to keep all of the terms when writing out (6.25) raised to the fifth power; we need only keep terms up to order α_2^2. That integral then proceeds with use of (6.31) and (6.32). To order α_2^2 and including a factor of 2π from integrating over ϕ, we have

$$
-\frac{2\pi}{3} \int\limits_{\theta,\phi} \int\limits_0^{r_1(\theta)} P_{0(1)} r_1^4 dr_1 d\Omega_1
$$

$$
= -\frac{4\pi^2 R_O^5}{15} \left\{ 2(1+\alpha_0)^5 + 4(1+\alpha_0)^3\alpha_2^2 + \dots \right\}. \quad (6.63)
$$

The second integral in (6.62) yields surviving terms that involve $(1+\alpha_0)^3$ and $(1+\alpha_0)\alpha_2^2$. When multiplying the result by the square-bracketed prefactor, again keep terms only up to order α_2^2. The result in this case is

$$
2\pi R_O^2 \left[(1+\alpha_0)^2 + \frac{1}{5}\alpha_2^2 \right] \int\limits_{\theta,\phi} \int\limits_0^{r_1(\theta)} P_{0(1)} r_1^2 dr_1 d\Omega_1
$$

$$
= \frac{8\pi^2 R_O^5}{3} \left\{ (1+\alpha_0)^5 + \frac{4}{5}(1+\alpha_0)^3\alpha_2^2 + \cdots \right\}. \quad (6.64)
$$

The third integral in (6.62) will again involve, $r_1^5(\theta)/5$, but here we need only keep terms up to order α_2 when expanding $r_1(\theta)$. This is because the prefactor in this case involves α_2 and α_2^2, so to keep terms *overall* to order α_2^2 we do not need to include the α_2^2 and higher-order terms when expanding. Only one term survives from this integral:

$$
\int\limits_{\theta,\phi} \int\limits_0^{r_1(\theta)} P_{2(1)} r_1^4 \, dr_1 d\Omega_1 = \frac{4\pi}{5} R_O^5 (1+\alpha_0)^4 \alpha_2. \tag{6.65}
$$

If (6.65) is multiplied by the prefactor in (6.62), the second term in the prefactor (the one involving [2, 2, 2]) would give rise to a term of order α_2^3, which we drop. Including the prefactor, the overall result for the last integral is then

$$
\frac{16\pi^2}{25} R_O^5 (1+\alpha_0)^3 \alpha_2^2 \tag{6.66}
$$

Gathering (6.63), (6.64), and (6.66) into (6.62) and simplifying gives

$$
U_C = \frac{4\pi \, \rho^2 R_O^5}{15 \, \varepsilon_0} \left\{ (1+\alpha_0)^5 + \frac{4}{5}(1+\alpha_0)^3 \alpha_2^2 + \cdots \right\}. \tag{6.67}
$$

On writing the charge density as $\rho = 3 Z e / 4\pi R_O^3$, again invoking $R_O \sim a_0 A^{1/3}$, and substituting the volume-conservation condition $\alpha_0 \sim -\alpha_2^2/5$, U_C reduces to

$$
U_C \sim a_C \left(\frac{Z^2}{A^{1/3}} \right) \left\{ 1 - \frac{1}{5}\alpha_2^2 + \cdots \right\}. \tag{6.68}
$$

where $a_C = \left(3 e^2 / 20 \pi \, \varepsilon_0 a_0 \right) \sim 0.72$ MeV is the Coulomb energy parameter. The Coulomb self-energy *decreases* upon perturbation of the nucleus from its initially spherical shape.

We can now determine the limiting condition for stability against spontaneous fission. If the nucleus becomes slightly distorted, that is, if $\alpha_2 \neq 0$, then fission will proceed spontaneously if the total energy of the deformed nucleus is less than what it was in its initial undeformed spherical shape ($\alpha_2 = 0$), that is, if $\Delta E = (U_S + U_C)_{deformed} - (U_S + U_C)_{undeformed} < 0$. On substituting (6.44) and (6.68), ΔE emerges as

$$
\Delta E = \left(\frac{2}{5} a_S A^{2/3} \alpha_2^2 \right) \left\{ 1 - \frac{1}{2}\left(\frac{a_C}{a_S} \right)\left(\frac{Z^2}{A} \right) \right\}. \tag{6.69}
$$

Clearly, whatever the value of α_2, ΔE will be negative so long as

$$\frac{Z^2}{A} > 2\left(\frac{a_S}{a_C}\right), \tag{6.70}$$

the Bohr and Wheeler SF condition! With $a_S \sim 18$ MeV and $a_C \sim 0.72$ MeV, the limiting Z^2/A evaluates to about 50. Readers seeking expressions for U_S and U_C to higher orders of perturbation are urged to consult Present and Knipp (1940a, b) and Plesset (1941).

With empirically-known values for a_S and a_C, the Z^2/A limit provides an understanding of why nature stocks the periodic table with only about 100 elements: nuclei have $A \sim 2Z$, so $Z^2/A \sim 50$ corresponds to a limiting Z of about 100. In extending their analysis to higher orders of perturbation, B&W also provided the first real understanding as to why only a very few isotopes at the heavy end of the periodic table are subject to fission by slow neutrons: yet heavier ones are too near the Z^2/A limit to remain stable for very long against SF, while for lighter ones the fission barrier is too great to be overcome by the binding energy released upon neutron capture.

6.6 Appendix F: Average Neutron Escape Probability from Within a Sphere

We derive here an expression for the mean escape probability for neutrons emitted from within a sphere, the quantity $\langle P_{sph} \rangle$ of Sects. 4.2 and 4.3. This is based on extending the semi-empirical one-dimensional expression

$$P(x) = \exp\left(-\sigma_{tot}\, n\, x\right) \tag{6.71}$$

to three dimensions. The approach taken here is adopted directly from Croft (1990).

Figure 6.6 shows an element of volume dV at radius r within a sphere of radius R. We can put this volume element somewhere along the z-axis without any loss of generality.

The vector $\boldsymbol{r}$ goes from the center of the sphere to dV, that is, $\boldsymbol{r} = r\,\hat{z}$. The vector $\boldsymbol{L}$ represents the straight-line path of a neutron emitted from dV in a direction that reaches the surface of the sphere, and $\boldsymbol{R} = \boldsymbol{r} + \boldsymbol{L}$ is the vector from the center of the sphere to where the neutron reaches the surface. $\boldsymbol{L}$ is directed at spherical-coordinate angles (θ, ϕ) measured from the location of dV; Fig. 6.7 shows a more detailed view of $\boldsymbol{L}$.

The approach to setting up an expression for $\langle P_{sph} \rangle$ comprises two parts. The first is to develop an expression for the probability that the path length L of a neutron's flight to the surface of the sphere lies between L and $L + dL$; this is designated as $P(L)dL$. This development is the lengthier of the two parts. Since the probability that a neutron that travels path length L to the surface will escape is $e^{-\sigma nL}$, then the probability that any one neutron will travel a path of length L to $L + dL$ *and* escape

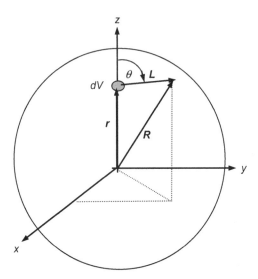

Fig. 6.6 Neutrons escaping from a small volume dV within a bomb core. A neutron begins at position r. R is its position when it reaches the surface of the core. Vector L, the neutron's straight-line path, goes from the volume element to the edge of the sphere in a direction defined by spherical-coordinate angles (θ, ϕ) as shown in Fig. 6.7

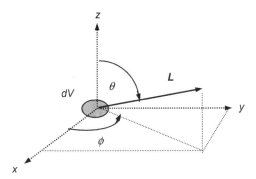

Fig. 6.7 Detailed view of vector L of Fig. 6.6

will be $e^{-\sigma n L}$ times $P(L)dL$. The second step is to integrate this product over all possible values of L to determine the average escape probability.

To begin the first step, apply the law of cosines to the triangle r-L-R:

$$R^2 = r^2 + L^2 + 2rL\cos\theta. \tag{6.72}$$

It is convenient to develop this derivation in terms of dimensionless variables. To this end, define

$$\rho = \frac{r}{R},$$

(6.73)

and

$$x = \frac{L}{R}.$$

(6.74)

ρ is referred to as the *reduced radial distance*, and x as the *reduced path length*. Also define

$$\mu = \cos \theta.$$

(6.75)

With these definitions, (6.72) can be written as

$$1 = \rho^2 + x^2 + 2\rho x \mu.$$

(6.76)

Solve this expression for μ:

$$\mu = \frac{1 - \rho^2 - x^2}{2\rho x}.$$

(6.77)

Now, compute the derivative of μ with respect to x, presuming that ρ is held constant. This gives

$$\left(\frac{\partial \mu}{\partial x} \right)_{\rho} = -\left[\frac{1}{\rho} + \frac{(1 - \rho^2 - x^2)}{2\rho x^2} \right].$$

(6.78)

This result will be valuable shortly. If we presume that neutrons are emitted homogeneously from within the entire sphere, then the probability $P(dV)$ that one will be emitted from within volume dV will be the ratio of dV to the volume of the entire sphere:

$$P(dV) = \frac{3}{4\pi R^3} dV.$$

(6.79)

If neutrons emitted from dV travel in random directions, then the probability that any one of them will be emitted into the solid angle defined by the angular limits θ to $\theta + d\theta$ and ϕ to $\phi + d\phi$ is

$$P(d\Omega) = \frac{1}{4\pi} \sin \theta \, d\theta \, d\phi.$$

(6.80)

The probability that a neutron will be emitted from dV into the direction $d\Omega$ will be the product of (6.79) and (6.80). In forming this product, replace $\sin\theta \, d\theta$ with

$-d\mu$ from (6.75):

$$P(dV, d\mu, d\phi) = -\frac{3}{16\pi^2 R^3} d\mu \, d\phi \, dV. \tag{6.81}$$

By using (6.78), we can transform (6.81) from an expression that gives the probability that a neutron will be emitted from within volume dV into the angular range $(d\phi, d\mu)$ into an expression in terms of an element of reduced path length dx:

$$P(dV, dx, d\phi) = +\frac{3}{16\pi^2 R^3} \left[\frac{1}{\rho} + \frac{(1 - \rho^2 - x^2)}{2\rho x^2} \right] dx \, d\phi \, dV. \tag{6.82}$$

The physical interpretation of this expression is that it gives the probability that a neutron will be emitted from within volume dV into a small range of azimuthal angle $d\phi$ in such a way that its reduced path length to the surface lies between x and $x + dx$. We can immediately integrate (6.82) over $\phi = 0$ to 2π to give

$$P(dV, dx) = +\frac{3}{8\pi R^3} \left[\frac{1}{\rho} + \frac{(1 - \rho^2 - x^2)}{2\rho x^2} \right] dx \, dV. \tag{6.83}$$

Now invoke a second set of spherical coordinates $\left(\theta', \phi' \right)$ measured with respect to the *origin* in Fig. 6.6. From the usual volume element in spherical coordinates and using (6.73), we can write any volume element dV as $dV = r^2 \sin\theta' dr \, d\theta' d\phi' = \rho^2 R^3 \sin\theta' d\rho \, d\theta' d\phi'$, and hence cast 6.83) as

$$P\left(d\rho, \, d\theta', \, d\phi', \, dx \right) = +\frac{3\rho^2}{8\pi} \left[\frac{1}{\rho} + \frac{(1 - \rho^2 - x^2)}{2\rho x^2} \right] \sin\theta' d\rho \, d\theta' d\phi' dx. \tag{6.84}$$

θ' and ϕ' can be integrated over directly; the result is 4π. Also, take one factor of ρ from outside the bracket in (6.84) to the inside, and bring the contents of the bracket to a common denominator. The result is

$$P(d\rho, dx) = \frac{3}{4}\rho \left[\frac{(1 - \rho^2 + x^2)}{x^2} \right] d\rho \, dx. \tag{6.85}$$

The physical interpretation of (6.85) is that it gives the probability of a neutron being emitted from within a shell of reduced radii from ρ to $\rho + d\rho$ in such a way that its reduced path length to the surface lies between x and $x + dx$. The next task is to integrate over ρ to transform this to an expression that gives purely the probability of a neutron's reduced path length lying between x and $x + dx$. To do this, we need to determine how the limits of integration over ρ depend on x.

The flight path length can vary from $L = 0$ to $2R$, so $0 \le x \le 2$. It is easiest to develop the relevant limits of integration in two regimes: $0 \le x \le 1$, and $1 \le x \le 2$.

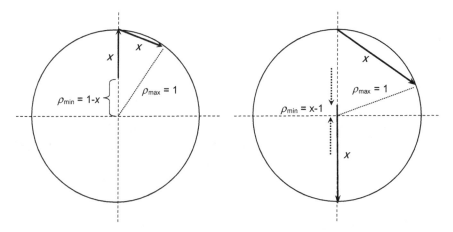

Fig. 6.8 Limits of integration over ρ for $0 \le x \le 1$ (left) and $1 \le x \le 2$ (right)

Refer to Fig. 6.8, where the sphere is drawn twice, imagined to be of reduced radius unity in both cases. The arrowed lines represent the length of x, which is ≤ 1 in the left panel, and between 1 and 2 in the right panel. In either case, if the flight path should happen to start from the top (or the bottom) of the sphere, we can always find an orientation for x such that it will also end at the edge of the sphere, where $\rho = 1$. This means that $\rho = 1$ is the *maximum* possible value of ρ for *any* value of x.

Look at Fig. 6.8 carefully. There are two arrowed lines in each diagram, one of which shows a starting place along the z-axis, and the other a starting place at the edge of the sphere. In the left figure, for which $0 \le x \le 1$, the *minimum* value of ρ that can be reached by the arrowed line occurs when it goes directly along the z-axis, in which case $\rho = 1-x$. On the other hand, in the right figure ($0 \le x \le 2$), a flight path that goes directly along the z-axis will have $\rho_{min} = x-1$. The arrowed lines can start anywhere, but *must* end at the edge of the sphere; you will have to *think* about this argument for a few minutes.

With these limits, we have

$$P_{0 \le x \le 1}(dx) = \int_{1-x}^{1} P(d\rho, dx) d\rho = \frac{3}{4x^2} \left\{ \int_{1-x}^{1} \left(\rho - \rho^3 + \rho x^2 \right) d\rho \right\} dx \quad (6.86)$$

and

$$P_{1 \le x \le 2}(dx) = \int_{x-1}^{1} P(d\rho, dx) d\rho = \frac{3}{4x^2} \left\{ \int_{x-1}^{1} \left(\rho - \rho^3 + \rho x^2 \right) d\rho \right\} dx. \quad (6.87)$$

These integrals give the same result:

$$P(dx) = \frac{3}{4}\left\{1 - \left(\frac{x}{2}\right)^2\right\}dx. \tag{6.88}$$

This expression is final result of the first part of the derivation of $\langle P_{sph}\rangle$: The probability that the reduced flight-path length lies between x and $x + dx$, or, equivalently, the probability that the true flight-path length in meters lies between L and $L + dL$.

The second part of the derivation is much shorter. If a neutron travels through flight path L from its birthplace to the edge of the sphere, then its probability of *not* being captured or causing a fission along the way is $\exp(-\sigma nL) = \exp(-\sigma nRx)$, where σ is the total (fission + capture) cross-section. To determine the overall average probability of escape, we need to integrate over all possible (reduced) flight-path lengths. For brevity, define $\Sigma = \sigma nR$. Then

$$\langle P_{sph}\rangle = \int_0^2 e^{-\Sigma x}P(dx) = \frac{3}{4}\int_0^2 e^{-\Sigma x}\left\{1 - \left(\frac{x}{2}\right)^2\right\}dx. \tag{6.89}$$

This integral is straightforward, and reduces to

$$\langle P_{sph}\rangle = \frac{3}{8\Sigma^3}\left\{2\Sigma^2 + e^{-2\Sigma}(2\Sigma + 1) - 1\right\}. \tag{6.90}$$

This is the expression used in Sects. 4.2 and 4.3 in dealing with predetonation and fizzle-yield probabilities.

6.7   Appendix G: The Neutron Diffusion Equation

The various analyses of criticality in Chap. 2 are predicated on the diffusion equation for neutrons, a differential equation for the time and space-dependence of the number density of neutrons within a bomb core. Fundamentally, the diffusion equation expresses a competition between neutron gain and loss. In some volume of interest within the core, neutrons will be gained both from fissions occurring within it and from those which enter from surrounding material. At the same time, the volume will lose neutrons as they are consumed in causing fissions and as they fly out into surrounding material (or to the outside world if the volume element should be at the edge of the core). The quantity of interest is the number density N of neutrons within the volume, which has units of neutrons per cubic meter and is presumed to be a function of both time and the location of the volume within the core. In anticipation of modeling a spherical core we write this as $N(r, t)$. In words, the net rate of change of neutron density can be expressed as

$$\frac{dN}{dt} = \begin{pmatrix} neutron\,density\,gain \\ from\,fissions \end{pmatrix}$$

$$+ \begin{pmatrix} rate\,of\,neutron\,density\,gain\,or\,loss \\ by\,transport\,through\,volume\,boundary \end{pmatrix}. \tag{6.91}$$

The derivation given here is motivated by that appearing in Serber (1992); readers seeking more details are urged to consult Liverhant (1960) or any similar text on reactor engineering. An excellent introduction to the basics of neutron diffusion can be found in Section. 12-4 of the popular *Feynman Lectures on Physics*; see Feynman et al. (1964).

We approach the development of the diffusion equation in two steps, each corresponding to one of the terms on the right side of (6.91). The density gain from fissions can be derived quite easily, so we examine that term first.

Assume that, on average, neutrons have speed $\langle v \rangle$. From the development in Sect. 2.1, we know that the average distance a neutron will travel before causing a fission is given by $\lambda_f = 1/n\sigma_f$, where n is the number density of fissile nuclei and σ_f is the fission cross-section. The time that a neutron will travel before causing a fission is then $\tau = \lambda_f / \langle v \rangle$. On average, individual neutrons will cause fissions at a rate $\langle v \rangle / \lambda_f$ per second. If each fission produces v secondary neutrons, then the net rate of secondary neutron production per "average" neutron will be $(v - 1)\langle v \rangle / \lambda_f$ per second; the "-1" appears because the neutron that causes the fission is consumed in doing so. Now apply this argument to a volume V where the number density of neutrons is N. The total number of neutrons will be NV, and the rate of secondary neutron production will consequently be $NV(v - 1)\langle v \rangle / \lambda_f$ per second. The rate of change of the density of neutrons caused by fissions is given by this quantity divided by V, or

$$\left(\frac{\partial N}{\partial t} \right)_{fission} = \frac{\langle v \rangle}{\lambda_f}(v - 1)\,N. \tag{6.92}$$

The second term in (6.91) involves neutrons entering and leaving the volume as they fly about. This step is trickier and is most easily dealt with in two sub-steps.

To begin, an important quantity here is the *transport mean free path*, the average distance a neutron will travel before suffering any interaction. In a bomb core the important interactions are fission and elastic scattering; again appealing to Sect. 2.1, we write this as

$$\lambda_t = \frac{1}{n\sigma_{total}} = \frac{1}{n\left(\sigma_{fission} + \sigma_{elastic}\right)}. \tag{6.93}$$

Imagine neutrons flying about in a spherical bomb core of radius R as sketched in Fig. 6.9. The first sub-step in this part of the derivation is to get an expression for the net rate at which neutrons flow from the inside to the outside through an imaginary surface at radius r.

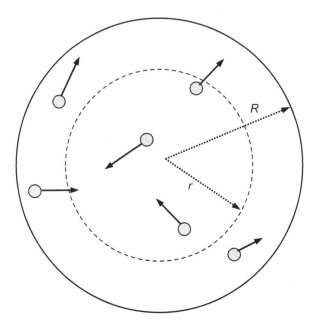

Fig. 6.9 Schematic representation of a fissioning spherical bomb core of radius R. The small filled circles represent neutrons. The neutron number density $N(r, t)$ is presumed to be a function both position and time within the core

This derivation makes use of a result established in Sect. 3.5, where we examined the effusion of particles through holes in a barrier. In deriving Eq. (3.54), we found that the effusion rate of particles through a hole of area A is given by

$$effusion\ rate = \frac{1}{4} N\ A\ \langle v \rangle. \tag{6.94}$$

The unit of this expression is neutrons per second, or, more compactly, sec^{-1}. In our case, A will be the area of the imaginary surface at radius r, that is, $4\pi r^2$.

Apply (6.94) to the imaginary surface at radius r. Unlike the barrier diffusion issue taken up in Sect. 3.5, here we have neutrons passing through the surface that have come from both "within" (radii $< r$) and "outside" (radii $> r$) the surface. Suppose that those neutrons which come from within arrive from a region where the average neutron number density is $N_<$, while those that pass through from the outside have come from a region where the neutron number density is $N_>$. The net neutron flux from the inside to the outside through the imaginary surface will then be

$$\begin{pmatrix} net\ effusion\ rate \\ inside\ to\ outside \\ at\ radius\ r \end{pmatrix} = \frac{1}{4} A\ \langle v \rangle\ (N_< - N_>). \tag{6.95}$$

While neutrons will on average travel distance λ_t of (6.93) between interactions, they will be flying about in random directions. In specifying the locations of $N_<$

and $N_>$, we should consequently use values corresponding to the average *radial* displacement that a neutron will undergo between its last collision and reaching the surface at r, that is, their average displacement perpendicular to the spherical surface. This will presumably be less than λ_t due to the neutrons' random motions. For the moment, let us represent this average radial displacement as $\langle \lambda_r \rangle$; how this is determined is taken up following (6.106) below.

Now, reverse the order of the terms in (6.95), and both multiply and divide by $2 \langle \lambda_r \rangle$:

$$\begin{pmatrix} net\ effusion\ rate \\ inside\ to\ outside \\ at\ radius\ r \end{pmatrix} = -\frac{1}{4} A \langle v \rangle \left[\frac{(N_> - N_<)}{2 \langle \lambda_r \rangle} \right] (2 \langle \lambda_r \rangle). \qquad (6.96)$$

The square bracket in (6.96) is the change in N divided by the distance over which that change occurs, that is, the derivative of N with respect to radial distance:

$$\begin{pmatrix} net\ effusion\ rate \\ inside\ to\ outside \\ at\ radius\ r \end{pmatrix} = -\frac{1}{2} A \langle v \rangle \langle \lambda_r \rangle \left(\frac{\partial N}{\partial r} \right)_r = -2 \pi r^2 \langle v \rangle \langle \lambda_r \rangle \left(\frac{\partial N}{\partial r} \right)_r,$$

$$(6.97)$$

where we have substituted for the area of the sphere and used partial derivatives as a reminder that N is a function of both position and time. Be sure to understand why factors of 2 were included with the factors of $\langle \lambda_r \rangle$ in (6.96): the surfaces of density $N_<$ and $N_>$ are each distance $\langle \lambda_r \rangle$ from the surface at radius r, and so the distance over which the change $(N_> - N_<)$ occurs is $2 \langle \lambda_r \rangle$.

We come now to the second sub-step of this part of the derivation. We desire an expression for the net rate of change of N per unit volume due to random neutron motions. To do this, apply (6.97) to a spherical shell within the core that extends from inner radius r to outer radius $r + dr$, as shown in Fig. 6.10. For neutrons arriving from inside the shell,

$$\begin{pmatrix} rate\ of\ neutrons\ from \\ within\ r\ entering\ shell \end{pmatrix} = -2 \pi r^2 \langle v \rangle \langle \lambda_r \rangle \left(\frac{\partial N}{\partial r} \right)_r. \qquad (6.98)$$

At the same time, neutrons exit the shell by passing through the surface at $r + dr$:

$$\begin{pmatrix} rate\ of\ neutrons\ exiting \\ shell\ from\ within \end{pmatrix} = -2 \pi (r + dr)^2 \langle v \rangle \langle \lambda_r \rangle \left(\frac{\partial N}{\partial r} \right)_{r+dr}. \qquad (6.99)$$

Notice that in writing these expressions, we evaluate $(\partial N / \partial r)$ at the inner and outer surfaces of the shell. It follows that the *net* rate of neutron flux into the shell is given by the entry rate, (6.98), minus the exit rate, (6.99); the overall result could in fact be a loss (and will be so at the outer surface of the core):

Fig. 6.10 Spherical shell of
material of inner radius r and
thickness dr

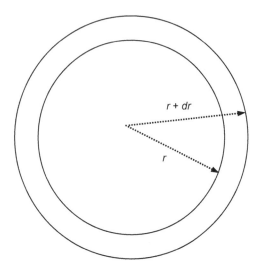

$$\begin{pmatrix} net\,rate \\ of\,neutrons \\ entering\,shell \end{pmatrix} = 2\,\pi\,\langle v \rangle\,\langle \lambda_r \rangle \left\{ (r+dr)^2 \left(\frac{\partial N}{\partial r} \right)_{r+dr} - r^2 \left(\frac{\partial N}{\partial r} \right)_r \right\}.$$

(6.100)

On expanding out the factor of $(r + dr)^2$ and writing

$$\left(\frac{\partial N}{\partial r} \right)_{r+dr} = \left(\frac{\partial N}{\partial r} \right)_r + \left(\frac{\partial^2 N}{\partial r^2} \right)_r dr,$$

(6.101)

one arrives at

$$\begin{pmatrix} net\,rate\,of\,neutrons \\ entering\,shell \end{pmatrix} = 2\,\pi\,\langle v \rangle\,\langle \lambda_r \rangle \left\{ \left[r^2 \left(\frac{\partial^2 N}{\partial r^2} \right)_r + 2r \left(\frac{\partial N}{\partial r} \right)_r \right] dr \right.$$
$$\left. +2r \left(\frac{\partial^2 N}{\partial r^2} \right)_r dr^2 + \left(\frac{\partial N}{\partial r} \right)_r dr^2 + \left(\frac{\partial^2 N}{\partial r^2} \right)_r dr^3 \right\}.$$

(6.102)

Now recall the Laplacian operator in spherical coordinates:

$$\left(\nabla^2 N \right)_r = \frac{1}{r^2} \left[r^2 \left(\frac{\partial^2 N}{\partial r^2} \right)_r + 2r \left(\frac{\partial N}{\partial r} \right)_r \right].$$

(6.103)

But for a factor of $1/r^2$, this is exactly the square-bracketed term in (6.102). Hence
we can write

$$\begin{pmatrix} net\ rate\ of \\ neutrons \\ entering\ shell \end{pmatrix} = 2\pi \langle v \rangle \langle \lambda_r \rangle \left\{ r^2 (\nabla^2 N)_r\, dr + 2r \left(\frac{\partial^2 N}{\partial r^2} \right)_r dr^2 + \right.$$

$$\left. \left(\frac{\partial N}{\partial r} \right)_r dr^2 + \left(\frac{\partial^2 N}{\partial r^2} \right)_r dr^3 \right\}. \quad (6.104)$$

Now, the volume of the shell is $4\pi r^2 dr$. If we divide (6.104) by this volume we will arrive at the rate of change of the density of neutrons within the volume due to neutrons flying into or out of it:

$$\left(\frac{\partial N}{\partial t} \right)_{\substack{neutron \\ flight}} = \frac{1}{2} \langle v \rangle \langle \lambda_r \rangle \left\{ (\nabla^2 N)_r + \frac{2}{r} \left(\frac{\partial^2 N}{\partial r^2} \right)_r dr + \frac{1}{r^2} \left(\frac{\partial N}{\partial r} \right)_r dr \right.$$

$$\left. + \frac{1}{r^2} \left(\frac{\partial^2 N}{\partial r^2} \right)_r dr^2 \right\}. \quad (6.105)$$

If we let the shell become infinitesimally thin, that is, if $dr \to 0$, then the last three terms on the right side of (6.105) will vanish and we are left with

$$\left(\frac{\partial N}{\partial t} \right)_{\substack{neutron \\ flight}} = \frac{1}{2} \langle v \rangle \langle \lambda_r \rangle (\nabla^2 N), \quad (6.106)$$

where we drop the subscript r on $\nabla^2 N$ for sake of brevity.

We now need to address the issue of expressing the average radial neutron-travel distance $\langle \lambda_r \rangle$ in terms of the transport cross-section of (6.93). To do this we again appeal to Sect. 3.5, where we looked at the rate of escape of particles from within a slanted "escape cylinder." From (3.52), the number of particles traveling in the range of spherical directions (θ, ϕ) to $(\theta + d\theta, \phi + d\phi)$ that escape in elapsed time Δt is given by

$$N_{esc}(\Delta t) = \frac{N A \langle v \rangle (\Delta t)}{4\pi} \cos\theta \, \sin\theta \, d\theta \, d\phi, \quad (6.107)$$

where N, A, and $\langle v \rangle$ are again respectively the neutron number density, the area of the surface of escape, and the average neutron speed.

In (6.107), $\langle v \rangle (\Delta t)$ corresponds to the average distance that a neutron travels while making its escape, that is, $\langle v \rangle (\Delta t) = \lambda_t$. Since θ is measured from the z-axis (review Figs. 3.9 and 3.11), the vertical component of this distance, that is, the average distance that a neutron travels in a direction perpendicular to the escape surface, will be $\lambda_t \cos\theta$. In the context of our spherical bomb core, this perpendicular direction translates into the distance that a neutron will travel in the *radial* direction while escaping, which is what we are after. The total radial distance traveled by neutrons that escape in time Δt will then be

$$
\begin{pmatrix}
total\,radial\,distance \\
traveled\,by\,all\,neutrons \\
moving\,in\,directions\,(\theta,\phi) \\
that\,escape\,in\,time\,\Delta t
\end{pmatrix}
=
\begin{pmatrix}
number\,that\,escape \\
in\,time\,\Delta t
\end{pmatrix}
$$

$$
\times
\begin{pmatrix}
average\,radial\,distance \\
traveled\,by\,each\,neutron
\end{pmatrix}
$$

$$
= \frac{N\,A\,\lambda_t^2}{4\pi}\ \cos^2\theta\ \sin\theta\ d\theta\ d\phi. \tag{6.108}
$$

To account for all possible direction of escape, integrate over $0 \le \theta \le \pi/2$ and $0 \le \phi \le 2\pi$:

$$
\begin{pmatrix}
total\,radial\,distance \\
traveled\,by\,all\,neutrons \\
that\,escape\,in\,time\,\Delta t
\end{pmatrix}
= \frac{N\,A\,\lambda_t^2}{4\pi} \int_{\phi=0}^{2\pi} \int_{\theta=0}^{\pi/2} \cos^2\theta\ \sin\theta\ d\theta\ d\phi. \tag{6.109}
$$

This double integral gives $2\pi/3$, so

$$
\begin{pmatrix}
total\,radial\,distance \\
traveled\,by\,all\,neutrons \\
that\,escape\,in\,time\,\Delta t
\end{pmatrix}
= \frac{N\,A\,\lambda_t^2}{6}. \tag{6.110}
$$

For use in (6.106), we need the *average* radial distance traveled. We can get this by dividing (6.110) by the total number of neutrons that escape in time Δt. Equation (6.94) gives the *rate* of escape (neutrons per second), so the number that escape in time Δt will just be that rate times Δt:

$$
\begin{pmatrix}
average\,radial\,distance \\
traveled\,by\,all\,neutrons \\
that\,escape\,in\,time\,\Delta t
\end{pmatrix}
= \frac{\left(\frac{N\,A\,\lambda_t^2}{6}\right)}{\left(\frac{1}{4}N\,A\,\langle v\rangle \Delta t\right)} = \frac{2}{3}\lambda_t, \tag{6.111}
$$

where we used $\langle v\rangle(\Delta t) = \lambda_t$. This result, when substituted into (6.106), gives

$$
\left(\frac{\partial N}{\partial t}\right)_{\substack{neutron \\ flight}} = \frac{1}{3}\,\langle v\rangle\,\lambda_t\left(\nabla^2 N\right). \tag{6.112}
$$

We have now established two important results. These are (i) That within a unit volume of core material, (6.92) accounts for the rate of change of neutron density caused by neutrons created by fissions, and, (ii), That (6.112) accounts for the change in density caused by neutrons entering or leaving the volume. The *total* rate of change of neutron density is the sum of these two effects:

$$\frac{dN}{dt} = \frac{v}{\lambda_f}\,(v-1)\,N + \frac{v\,\lambda_t}{3}\,(\nabla^2 N), \tag{6.113}$$

where we have dropped the angle brackets on the average neutron speed. This is the diffusion equation used in Sect. 2.3 to study critical mass.

Solving (6.113) can be approached by the usual separation-of-variables technique. To actually determine a critical radius R, however, requires a boundary condition, that is, some constraint on $N(R)$. Establishing this condition requires being a little more careful with our derivation in (6.95) and (6.96) when applied to the edge of the sphere. Consider first (6.95) applied to the surface of the sphere at radius R. Here there will be no "backflow" of neutrons from the outside; the only neutrons that pass through the surface of the core will be those which have come from a characteristic distance $\langle \lambda_r \rangle$ from within. In this case, (6.95) reduces to

$$\begin{pmatrix} net\ effusion\ rate \\ through\ core\ surface \end{pmatrix} = \frac{1}{4} A\,\langle v \rangle\,(N_<). \tag{6.114}$$

Now consider (6.96) at the surface. The role of $N_>$ will be played by N_R, that is, the neutron density at the surface. In this case we have the change in N over only a distance of $\langle \lambda_r \rangle$ as opposed to the previous $2\langle \lambda_r \rangle$ since there is no inwardly-directed flux from the outside:

$$\begin{pmatrix} net\ effusion\ rate \\ through\ core\ surface \end{pmatrix} = -\frac{1}{4} A\,\langle v \rangle \left[\frac{(N_R - N_<)}{\langle \lambda_r \rangle} \right] \langle \lambda_r \rangle$$

$$= -\frac{1}{4} A\,\langle v \rangle\,\langle \lambda_r \rangle \left(\frac{\partial N}{\partial r} \right)_R. \tag{6.115}$$

Now demand consistency between (6.114) and (6.115); also invoke (6.111) for $\langle \lambda_r \rangle$. On approximating $N_< \sim N_R$ in (6.114), we find

$$N(R) = -\frac{2}{3} \lambda_t \left(\frac{dN}{dr} \right)_R. \tag{6.116}$$

This is the boundary condition used in Sect. 2.2 for determining critical mass.

It is important to point out that a diffusion approach to calculating critical mass contains some inherent level of approximation. In (6.111), it is determined that the average radial distance traveled by neutrons as they escape the core is $2\lambda_t/3$. If the computed core size should prove to be not much larger than this figure, one has to question the meaning of such an average. From the figures given in Table 2.1, $2\lambda_t/3$ ~ 2.4 cm for ^{235}U and 2.7 cm for ^{239}Pu. In comparison, the computed critical radii are 8.4 and 6.3 cm, which are about 3.5 and 2.3 times the average radial path lengths. Our result for the critical mass of ^{239}Pu might thus in particular be regarded with some skepticism. This issue is discussed further in Sect. 2.2, where it is pointed out that, despite this complication, a diffusion model predicts critical masses in good accord with experimentally-measured values.

6.8 Appendix H: Exercises and Answers

1.1 Compute Q-values for the following reactions. Reaction (a) produces high-energy neutrons for use in so-called "boosted" fission weapons. Reaction (b) is important in the production of tritium for use in reaction (a). Reaction (c) is a hypothetical fission reaction. Reaction (d) is an example of how alpha-bombardment of a light element can release neutrons, an important consideration in avoiding pre-detonation in fission weapons.

(a) $^{2}_{1}H + ^{3}_{1}H \rightarrow ^{4}_{2}He + ^{1}_{0}n$
(b) $^{6}_{3}Li + ^{1}_{0}n \rightarrow ^{3}_{1}H + ^{4}_{2}He$
(c) $^{1}_{0}n + ^{238}_{92}U \rightarrow 2\left(^{118}_{46}Pd\right) + 3\left(^{1}_{0}n\right)$
(d) $^{4}_{2}He + ^{27}_{13}Al \rightarrow ^{30}_{15}P + ^{1}_{0}n$

1.2 To melt one gram of ice at 0 C into 1 g of water at 0 C requires input of 80 physical calories of heat energy. If all of the energy involved in the alpha-decay of one gram ^{226}Ra over the course of one day could directed into melting ice, how many grams of ice could be melted per day? The decay rate of ^{226}Ra is 3.7 $\times 10^{10}$ per gram per second, and the alphas have kinetic energies of 4.8 MeV.

1.3 Prove Eq. (1.20) (assume classical mechanics), and then apply it to the case of radium decay discussed in Sect. 1.2. What will be the kinetic energy of the emergent α-particle? How does your result compare to the value of 4.78 MeV quoted in the *Chart of the Nuclides*?

1.4 For each of the reactions below, compute the energy of the resulting γ-ray, assuming that it is moving forward after the reaction. Assume that the target nucleus is stationary in each case.

(a) $^{1}_{1}H + ^{16}_{8}O \rightarrow ^{17}_{9}F + \gamma$ $(K_H = 4.9 \text{ MeV})$
(b) $^{4}_{2}He + ^{27}_{13}Al \rightarrow ^{31}_{15}P + \gamma$ $(K_{He} = 6.5 \text{ MeV})$
(c) $^{56}_{26}Fe + ^{94}_{40}Zr \rightarrow ^{150}_{66}Dy + \gamma$ $(K_{Fe} = 50 \text{ MeV})$

1.5 For each of the two-body reactions below, compute the Q-value of the reaction, the threshold energy (if any), and the kinetic energies and directions of motion of the products. Assume that the target nucleus is initially stationary in each case.

(a) $^{4}_{2}He + ^{27}_{13}Al \rightarrow ^{30}_{15}P + ^{1}_{0}n$ $(K_{He} = 5 \text{ MeV})$
(b) $^{2}_{1}H + ^{3}_{1}H \rightarrow ^{4}_{2}He + ^{1}_{0}n$ $(K_H = 3 \text{ MeV})$
(c) $^{4}_{2}He + ^{19}_{9}F \rightarrow ^{22}_{10}Ne + ^{1}_{1}H$ $(K_{He} = 2.75 \text{ MeV})$
(d) $^{4}_{2}He + ^{56}_{26}Fe \rightarrow ^{35}_{16}S + ^{25}_{12}Mg$ $(K_{He} = 30 \text{ MeV})$
(e) $^{16}_{8}O + ^{238}_{92}U \rightarrow ^{252}_{99}Es + ^{2}_{1}H$ $(K_O = 60 \text{ MeV})$

1.6   Consider a γ-ray of energy E_γ and a classical, non-relativistic particle of mass m moving with the same amount of kinetic energy. Both strike a classical, non-relativistic particle of mass M head on as in Sect. 1.4; assume that the gamma recoils backwards. Show that the ratio of the kinetic energy acquired by M when struck by the massive particle to that when struck by the γ-ray is approximately

$$\frac{K_M^m}{K_M^\gamma} \sim \frac{2E_m}{E_\gamma \left(1 + E_m / E_M\right)^2},$$

where the E's designate mc^2 rest energies and where it has been assumed that $E_\gamma << E_M$. Apply to an α-particle being struck by a γ-ray and a proton, with $E_\gamma = 10$ MeV. HINT: Consider Eqs. (1.32) and (1.34).

1.7   Show that the kinetic energy of a nonrelativistic neutron moving with speed $v = \beta c$ is given by $E \sim 470\,\beta^2$ MeV.

1.8   In an environment of absolute temperature T, the kinetic energy of a particle corresponds on average to $3k_B T/2$, where k_B is Boltzmann's constant. Show that if a neutron is moving at a nonrelativistic speed with kinetic energy E MeV, then the equivalent temperature is $T = (7.74 \times 10^9\,E)$ Kelvin. Energies of a couple MeV are characteristic of neutrons released in fission reactions.

1.9   See Fig. 6.11. A neutron of mass m and kinetic energy K (non-relativistic) strikes and is captured by a heavy nucleus of mass $2\,M >> m$. The resulting compound nucleus flies off with kinetic energy K_C. Shortly thereafter, the compound nucleus fissions into two equal halves, each of mass M. One fragment travels backward with kinetic energy K_B while the other continues forward with kinetic energy K_F. Energy $2Q$ is liberated in the fission, that is, $K_B + K_F - K_C = 2Q$. Show that the difference in kinetic energies between the forward and backward-moving fission fragments $\Delta Q = K_F - K_B$ is given to a good approximation by

$$\frac{\Delta Q}{Q} \sim 2\sqrt{\frac{Km}{QM}}.$$

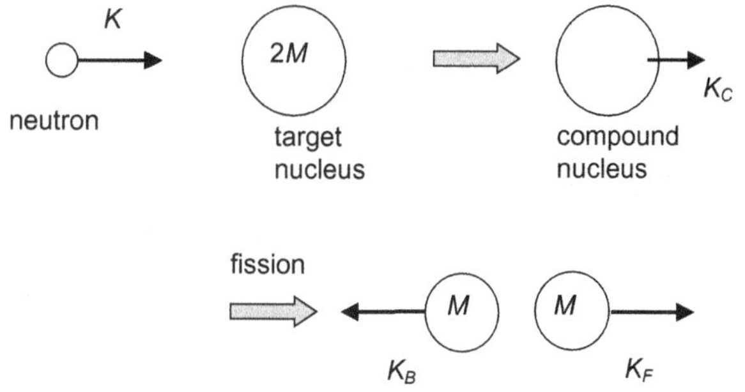

Fig. 6.11 Problem 1.9

Fig. 6.12 Problem 1.12

Apply your result to a neutron with $K = 14$ MeV striking a ^{235}U nucleus; what is $\Delta Q/Q$ if $Q = 100$ MeV? HINTS: Conserve momentum in each reaction. Assume that $M >> m$, $K << 2Q$, and that $\Delta Q/K_B$ (or $\Delta Q/K_A$) is small. Are these approximations justified in the 14-MeV neutron $+$ ^{235}U reaction?

1.10 Suppose that all of the energy liberated in the explosion of a 20-kiloton fission weapon could be directed into raising 1 cubic kilometer of water in the Earth's gravitational field. How high could that cubic km of water be raised?

1.11 What does the calibration of Eq. (1.89) predict for the fission barrier energy of ^{236}U? How does this compare to the value given in Appendix A? Does the model predict that this nuclide would be fissile for the Q-value discussed in Sect. 1.9?

1.12 See Fig. 6.12. A nucleus containing Z_1 protons approaches a fixed target nucleus containing Z_2 protons and a total of A nucleons; the kinetic energy of the incoming nucleus is K MeV when it is far from the target nucleus. If nuclear radii are described empirically by $R \sim a_O A^{1/3}$ where $a_O = 1.2$ fm, show that the ratio of the distance d of closest approach of the nuclear centers to the radius of the target nucleus is given by

$$\frac{d}{R} = 1.2\left(\frac{Z_1 Z_2}{K A^{1/3}}\right).$$

Apply to an alpha-particle with $K = 5$ MeV approaching a ^{235}U nucleus.

1.13 According to quantum physics, a particle of mass m moving with kinetic energy K has a wave nature, with a *de Broglie wavelength* given by $\lambda = h\big/\sqrt{2mK}$, where h is Planck's constant. If λ is on the order of or greater than the size of a target particle that is struck by the moving particle, then the collision cannot be analyzed with ordinary kinematics because we really have no idea of the geometry of the collision. Consider a neutron striking a nucleus of mass number A. Using the empirical nuclear-radius expression adopted in the previous problem, show that if we set the definition of a particle-like interaction to be that λ must be less than or approximately equal to the diameter of the struck nucleus, then the necessary kinetic energy must satisfy

$$K \gtrsim \frac{142}{A^{2/3}} \text{ MeV}.$$

Is this satisfied in the case of a 1-MeV neutron striking a nucleus of ^{235}U?

2.1 Using the empirical nuclear-radius expression of the previous problems, estimate the geometrical cross-sectional area of a ^{235}U nucleus; give your result
in barns. How does your result compare to the fission cross-section for fast
neutrons for this isotope, $\sigma_f = 1.235$ bn?

2.2 Because cadmium-113 has an enormous cross-section for capturing thermal
neutrons, strips of cadmium metal are often used in control mechanisms in
reactors. Given $\rho = 8.65$ gr cm^{-3}, $A = 112.904$ gr mol^{-1} and $\sigma_{capture} =$
20,600 barns, compute the probability that a neutron will penetrate through a
strip of Cd-113 of thickness 0.05 mm.

2.3 Neutrons with speed v_{neut} corresponding to classical kinetic energy K MeV are
traveling between nuclei in a material of atomic weight A gr mol^{-1}, density
ρ gr cm^{-3}, and fission cross-section σ_f barns. Show that the time between
fissions $\tau = \lambda_f / v_{neut}$ can be expressed as

$$\tau = \frac{1.20 \, A}{\sigma_f \, \rho \, \sqrt{K}} \quad \text{nanoseconds}$$

Compute τ for the case of 2-MeV neutrons in ^{235}U: $A = 235$, $\rho = 18.71$ gr
cm^{-3} and $\sigma_f = 1.235$ bn.

2.4 For ^{233}U, $n = 4.794 \times 10^{22}$ cm^{-3}, $\sigma_f = 1.946$ b, $\sigma_{el} = 4.447$ b, $v = 2.755$, and
$\rho = 18.55$ g cm^{-3}. What does diffusion theory indicate for the critical mass
of this isotope? ^{233}U would make an excellent nuclear explosive, but it does
not occur naturally.

2.5 You have 5 kg of beryllium-oxide to use as a tamper. What uncompressed
mass of ^{239}Pu can be rendered just critical, assuming spherical geometry?

2.6 Working from the development of bomb efficiency in Sect. 2.4, show that the
speed of the expanding core at the time of second criticality is given by

$$v(t_{crit}) = \frac{\alpha \, \Delta r}{\tau}.$$

Evaluate $v(t_{crit})$ for a ^{235}U core with $C = 1.5$ bare critical masses using the
values given in Table 2.2. How does $v(t_{crit})$ compare to the average neutron
speed?

2.7 Working from the development of bomb efficiency in Sect. 2.5, show that the
pressure within the expanding core at the time of second criticality is given by

$$P(t_{crit}) = \frac{\alpha^2 \, \Delta r \, \langle \rho r \rangle}{3 \, \tau^2}.$$

Evaluate $P(t_{crit})$ in the case of a ^{235}U core with $C = 1.5$. Express your result
in atmospheres: 1 atm $\sim$ 101,000 Pa.

2.8 Consider a hypothetical fissile material with $\rho = 17.3$ gr cm^{-3}, $A = 250$ gr
mol^{-1}, $\sigma_f = 1.55$ bn, $\sigma_{el} = 6$ bn, and $v = 2.95$. Use **CriticalityAnalytic**

to determine the bare threshold (spherical) critical radius and mass of this material. What are the values of α, Δr, efficiency, and yield for a core of $C = 3$ critical masses of this material if $E_f = 185$ MeV and if the secondary neutrons have $E = 2$ MeV? Take $\gamma = 1/3$. If the initial number of neutrons is taken to be one, what are the fission and criticality-shutdown timescales? (Do not worry about any refinements to the efficiency calculation as discussed at the end of Sect. 2.6.) If you have a FORTRAN compiler available, obtain the numerical-simulation program of Sect. 2.6 and apply it to the same situation; what yield does the program predict?

2.9   What outer radius and mass of tungsten-carbide tamper will render just threshold critical a 7-kg spherical core of uncompressed ^{239}Pu? Hint: Fig. 2.10.

2.10 Consider a mass m of a pure fissile material whose normal density is ρ_o, with m being less than the bare threshold critical mass for the material. Show that this mass can be made critical by compressing it to a radius given by

$$r_{compress} \leq \sqrt{\frac{3m}{4\pi\rho_o R_o}},$$

where R_o is the threshold critical radius at normal density. Show further that if C is the number of threshold critical masses represented by m ($C < 1$), then the ratio of the density at this compressed radius to the initial density is given by

$$\frac{\rho_{compress}}{\rho_o} = \sqrt{\frac{1}{C}}.$$

Evaluate numerically for 100 grams of ^{235}U.

2.11 The diffusion equation for neutrons in a bomb core, (2.18), can be applied in any coordinate system, provided that the expression for $\nabla^2 N$ in that system is used. To this end, consider a *cubical* core that extends from $0 \leq x \leq L$, $0 \leq y \leq L$, and $0 \leq z \leq L$. Solve the diffusion equation in Cartesian coordinates. Show that if the simplified boundary condition $N(L_C) = 0$ is used, then the side length for threshold criticality is given by

$$L_C = \frac{\pi}{\sqrt{\nu - 1}}\sqrt{\lambda_f \lambda_t},$$

where the symbols have the same meanings as in Sect. 2.2. Compare this result to that in Sect. 2.8.1 to show that the critical mass for a cubical bomb core of a given material is $3^{5/2}/(4\pi) \sim 1.24$ times that of a spherical core of the same material with this boundary condition. If you are familiar with quantum mechanics, the solution to this problem is very similar to that of a particle in a three-dimensional infinite potential box.

2.12 If you have a FORTRAN compiler available, obtain the numerical-simulation program of Sect. 2.6 and apply it to the following situation. A fission bomb is to be made of a pure ^{239}Pu core (normal density) of mass 30 kg plus a ^{238}U tamper of mass 200 kg ($A = 238.05$ gr mol^{-1}; $\rho = 18.95$ gr cm^{-3}; $\sigma_{el} = 4.804$ bn; ignore any fissility of the tamper). What is the yield? Take $\gamma = 1/3$, the average neutron energy to be 2 MeV, the energy per fission to be 180 MeV, and assume one initiating neutron.

2.13 Repeat the derivation of Sect. 2.3 for threshold criticality of a tamped core ($\alpha = \delta = 0$), but take the "outer" boundary condition to be simplified to $N_{tamp}(R_{tamp}) = 0$. Show that the criticality condition simplifies to

$$1 - \left(R_{thresh}/d_{core}\right) \cot\left(R_{thresh}/d_{core}\right) = \frac{\left(\lambda_{trans}^{tamp}/\lambda_{trans}^{core}\right)}{\left(1 - R_{thresh}/R_{tamp}\right)}.$$

In this approximation, what is the threshold critical radius for ^{235}U for a tungsten-carbide tamper of outer radius 16 cm? Compare to the result for an untamped core with the same boundary condition, Sect. 2.1.8.

2.14 It is remarked in Sect. 2.5 that in the case of a gas of uranium nuclei of normal density of that metal (18.95 gr cm^{-3}), radiation pressure dominates gas pressure for per-particle energies greater than about 2 keV. This problem investigates this issue.

For a "gas" of photons, thermodynamics provides the following expression for the pressure:

$$P_{rad} = \left(\frac{8\pi^5 k_B^4}{45c^3 h^3}\right) T^4,$$

where k_B is Boltzmann's constant, c is the speed of light, h is Planck's constant, and T is the absolute temperature. For a gas of "classical" particles, the ideal gas law can be cast as

$$P_{classical} = 10^6 \left(\frac{\rho N_A k_B}{A}\right) T,$$

where ρ and A are the density and atomic weight of the material in gr cm^{-3} and gr mol^{-1}, respectively; N_A is Avogadro's number as usual. Given that in the classical case the per-particle energy is $3k_B T/2$, show that radiation pressure will dominate over gas pressure for classical per-particle energies satisfying

$$E > \frac{3}{2}\left(\frac{45 \times 10^6}{8\pi^5}\right)^{1/3} \left(\frac{ch}{e}\right) \left(\frac{\rho N_A}{A}\right)^{1/3}$$

$$\sim \left(4.908 \times 10^{-5}\right) \left(\frac{\rho N_A}{A}\right)^{1/3} \text{eV},$$

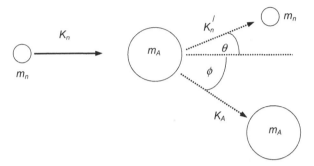

Fig. 6.13 Problem 3.1

where e is the electron charge. Hence verify the ~2 keV figure for uranium.

2.15 Suppose that the elastic-scattering cross-section, density, number of neutrons per fission, and atomic weight of ^{235}U are held fixed at the values used in Table 2.1, but that you can vary the fission cross-section σ_f from, say, 0.3 to 1.7 barns. Examine the trend of the bare threshold critical mass as a function of σ_f and determine if some simple empirical power-law dependence holds.

3.1 See Fig. 6.13. A non-relativistic neutron initially traveling in the x-direction with kinetic energy K_n suffers a *completely elastic* collision with an initially stationary nucleus of rest mass m_A. The neutron scatters through angle θ and the struck nucleus scatters through angle ϕ as shown. After the collision, the neutron and struck nucleus have kinetic energies K'_n and K_A, respectively. By conserving classical kinetic energy and momentum, eliminate ϕ and K_A to show that the initial and final neutron kinetic energies are related as

$$\sqrt{\frac{K'_n}{K_n}} = \frac{\cos\theta + \sqrt{\cos^2\theta + A^2 - 1}}{A + 1},$$

where A is the mass ratio m_A/m_n. If a neutron strikes an initially stationary carbon nucleus ($A = 12$) and scatters through $\theta = 90°$, what will be the its final speed in terms of its initial speed? Compare to the head-on case examined in Sect. 3.2.

3.2 Consider an ideal gas trapped within a sealed container at absolute temperature T. Working from the development in Sect. 3.5, show that the number of molecules that strike a square-meter area of the container wall per second is given by

$$\frac{P}{\sqrt{2\pi m k_B T}},$$

where P is the pressure, m is the mass of an individual molecule, and k_B is Boltzmann's constant. HINT: Use the Boltzmann's-constant form of the

Ideal Gas Law (Exercise 2.14). Air is mostly diatomic nitrogen (N_2); standard atmospheric pressure is about 101,000 Pa. Evaluate your answer for $T = 300$ K for air.

3.2 Consider the first stage of a gaseous diffusion plant for enriching uranium, where essentially all (139/140) of the atoms are ^{238}U. Suppose that vaporized pure uranium at $T = 300$ K and $P = 1$ atmosphere is pumped against a barrier; assume a vacuum on the other side. Working from your result in the previous problem, what total "hole area" S will you need if you want to process 140 kg of uranium per day? This would correspond to processing (although not isolating) 1 kg ^{235}U per day.

3.4 As in Exercise 3.2, consider an ideal gas trapped within an initially sealed container at absolute temperature T. The wall of the container is punctured, resulting in a small hole of area A through which molecules of the gas can effuse; assume that the outside environment is a vacuum so that nothing effuses back to the inside. Effusion represents a net *loss* of molecules from within the container.

(a) Working from the development in Sect. 3.5, show that, as a function of time, the pressure within the container will behave as

$$P = P_o e^{-t/\tau},$$

where P_o is the initial pressure and τ is a *characteristic effusion timescale* given by

$$\tau = \frac{4V}{A \langle v \rangle},$$

where V is the volume of the container and $\langle v \rangle$ is the average molecular speed. Assume that the temperature inside stays constant. The meaning of τ is that if the hole is not plugged, the pressure will drop to $1/e \sim 0.37$ of its initial value after τ seconds.

(b) A spacecraft cabin of volume 5 m^3 is punctured by a meteor, resulting in a hole of area 1 cm^2. If you model the atmosphere inside as pure diatomic nitrogen initially at standard atmospheric pressure and $T = 300$ K, what is the timescale τ in this case? Average molecular speed as a function of temperature is given by

$$\langle v \rangle = \sqrt{\frac{8 k_B T}{\pi m}}.$$

3.5 A reactor fueled with uranium enriched to $F = 0.06$ produces electrical power at a rate of 750 MW with a efficiency $\eta = 0.29$. What will be the rate of plutonium production in this reactor? Take $\sigma_{f5} = 584$ bn, $\sigma_{c8} = 2.7$ bn, and a fission energy of 180 MeV per reaction.

3.6   A news story indicates that a country that is attempting to develop nuclear weapons has succeeded in enriching 1000 kg of uranium hexafluoride to 5%, that is, 5% of the uranium nuclei in the sample are ^{235}U while the other 95% are ^{238}U. How many grams of ^{235}U are contained in the sample?

4.1. The half-life for spontaneous fission of $^{242}_{96}$Cm is 7.0×10^6 yr. What is the corresponding rate of spontaneous fissions per kg per second?

4.2. In the study of thermodynamic properties of materials, the following simple differential equation is used to model the change in volume dV of a sample of material of initial volume V when it is subjected to a change in pressure dP:

$$\frac{dV}{dP} = -\frac{V}{B}.$$

B is the *bulk modulus* of the material, a measure of its compressibility; a higher-B material is more difficult to compress than one of lower B. Assume that over the range of some compressive event, B averages 30 GPa. If an implosion bomb subjects a plutonium core to a pressure increase of one million atmospheres $(1 \text{ atm} \sim 10^5 \text{ Pa})$, integrate the differential equation to estimate the ratio of the final volume of the plutonium to its initial volume.

4.3   A rogue militia organization in an unstable country claims to have acquired 20 kg of ^{239}Pu of normal density and to have developed a crude gun-type bomb. The core contains 3% Pu 240. If they are to have a 50–50 chance of non-predetonation, what is the maximum tolerable assembly time? Take $S_{max} = 0$.

4.4   A 5-kg ^{239}Pu core is compressed to a density of 20 gr cm^{-3} at the time it reaches first criticality. If it is contaminated with 1.8% ^{240}Pu, what is the probability of achieving 70% of the design yield? Take $(\tau, F, \alpha_O, t_O) = (10^{-8} \text{ s}, 45, 1, 10^{-5} \text{ s})$ as in Sect. 4.3. What value of t_{init} corresponds to these conditions?

4.5   Prove that Eq. (4.21) reduces to Eq. (4.18) for $t_{init} \to 0$.

4.6   According to the publication of West & Sherwood cited in Sect. 4.4, the yield of 5.2 MeV alphas on ^{27}Al is 4.25×10^{-7}. If the number-density ratio of aluminum to the fissile material in a bomb core is held to 10^{-5}, what maximum rate of alpha-decays can be tolerated if the production of neutrons is to be kept to no more than 10^4 per second?

5.1   The purpose of this problem is to make a very crude estimate of the radioactivity produced by a fission weapon.
      Suppose that fission of ^{235}U happens exclusively by the reaction

$$^{235}_{92}\text{U} + ^1_0\text{n} \rightarrow ^{141}_{56}\text{Ba} + ^{92}_{36}\text{Kr} + 3\left(^1_0\text{n}\right).$$

Assume that 1 kg of ^{235}U is fissioned in this way. ^{141}Ba and ^{92}Kr both subsequently decay by beta-decay with half-lives of 18 min and 1.8 s, respectively.

Use the decay rate expression of Sect. 4.2 to estimate the "immediate" beta-radioactivity so generated; for sake of simplicity, ignore the neutrons released in the above reaction. If this radioactivity falls out over an area of 10 square miles, what will be the resulting immediate radioactivity in Curies per square meter? To put your result in perspective, household smoke detectors use 1 μCi alpha-emitters as ionization sources to help detect smoke particles.

F.1 In Appendix F, an expression is derived for the probability that the reduced flight-path-length x ($0 \le x \le 2$) for a neutron escaping from within a sphere will lie between x and $x + dx$:

$$P(dx) = \frac{3}{4}\left\{1 - \left(\frac{x}{2}\right)^2\right\}dx.$$

(a) What is the average value of the reduced escape-path length, $\langle x \rangle = \int x P(dx)$? (b) Determine also the root-mean-square escape-path length, $\sqrt{\langle x^2 \rangle}$. (c) A convenient measure of the "scatter" of measured x-values is the quantity $\Delta x = \sqrt{\langle x^2 \rangle - \langle x \rangle^2}$. What is Δx here?

Answers

1.1 (a) 17.59 MeV, (b) 4.78 MeV, (c) 182.0 MeV, (d) –2.642 MeV
1.2 7.34 grams
1.3 4.784 MeV
1.4 (a) $E_\gamma = 5.24$ MeV (b) $E_\gamma = 15.44$ MeV (c) Reaction impossible; $E_\gamma = -48$ MeV
1.5 (a) Threshold 3.034; $Q = -2.642$; P & n energies 0.307, 2.051; both forward
     (b) No threshold; $Q = 17.59$; He & n energies 0.99, 19.60; He backward; n forward
     (c) No threshold; $Q = 1.674$; Ne & H energies 0.068 and 4.356; both forward
     (d) Threshold 17.298; $Q = -16.143$; S & Mg energies 1.304 and 12.55; S backward; Mg forward
     (e) Threshold 51.07; $Q = -47.85$; Es & D energies 2.816 and 9.331; both forward
1.6 $K^{proton}_{alpha} / K^{\gamma}_{alpha} \sim 120$
1.9 $\Delta Q/Q \sim 0.07$; all approximations satisfied
1.10 8.57 meters
1.11 5.47 MeV; fissile
1.12 $d/R = 7.16$; $K \ge 3.73$ MeV; not satisfied
2.1 $\sigma \sim 1.723$ bn
2.2 Penetration probability ~0.00863
2.3 8.63 ns
2.4 14.2 kg
2.5 ~8.6 kg
2.6 $v(t_{crit}) \sim 2.38 \times 10^5$ m s^{-1}; ~1.2% neutron speed

2.7 $P(t_{crit}) \sim 4.73 \times 10^{15}$ Pa ~ 47 billion atmospheres

· 2.8 $R_o = 6.74$ cm; $M_o = 22.2$ kg. For $C = 3$, $\alpha = 0.841$, $\Delta r = 1.95$ cm, efficiency $= 25.4\%$, yield $= 280$ kt (at 16.5 kt/kg), $t_{fiss} = 0.57$ μs, $t_{crit} = 0.55$ μs. The numerical simulation program predicts a yield of ~5.46 kt. This is about 1/50 of the analytically-predicted yield, similar to the situation discussed at the end of Sect. 2.6

2.9 16.22 cm; 272 kg

2.10 $r_{compress} = 3.90$ mm; $\rho_{compress} = 21.42 \, \rho_o$

2.11 Yield ~ 6.9 kt. Initial core radius is 7.71 cm; core radius at second criticality ~9.24 cm. For this 200 kg tamper, the threshold-critical core mass is ~11.1 kg.

2.12 $R_{thresh} = 6.57$ cm, $M_{thresh} = 22.3$ kg

2.15 To a good approximation, $M \sim \sigma_f^{-2.05}$

3.1. $v_{final}/v_{initial} = \sqrt{143}/13 \sim 0.92$

3.2. Strike rate ~ 2.904×10^{27} m^{-2} s^{-1}

3.3 4.12×10^{-6} m^2

3.4 About 7 min

3.5 223 gr/day

3.6 33.395 grams

4.1 7.806×10^9 kg^{-1} s^{-1}

4.2 $V_{final}/V_{initial} \sim 0.036$

4.3 2.70 μs

4.4 74%; $t_{init} = 8.355$ μs

4.6 7.0×10^{15} s^{-1}

5.1 1.04×10^6 Ci m^{-2}

F.1 (a) 3/4, (b) 0.894, (c) 0.487

6.9 Appendix I: Glossary of Symbols

The great number of physical constants and parameters used in this book make it inevitable that different quantities in different chapters/sections become designated by the same symbol. The most important definitions are listed here, with defining equation numbers (in brackets) and prominent-use section numbers as appropriate. Symbols used in very localized contexts are not listed, nor are common usages such as (x, y, z) to designate Cartesian coordinates. English letters are listed first, followed by Greeks.

English letters

A mass number; nucleon number; constant of integration in electromagnetic separation of isotopes analysis (Sect. 3.4); sometimes used as surface area

A_H mass number of heavy-element fissile material in impurity calculation (Sect. 4.4)

A_L mass number of light-element contaminant in impurity calculation (Sect. 4.4)

a_S	surface-energy parameter in fission analysis (~ 18 MeV)
a_C	Coulomb-energy parameter in fission analysis (~ 0.72 MeV)
a_O	empirical parameter involved in nuclear sizes (~ 1.2 fm)
B	parameter in analysis of fission barrier (1.89); magnetic field strength (Sect. 3.4); boron to carbon number ratio (Sect. 4.1); bulk modulus (Sect. 4.2)
b	astronomical brightness; Sect. 5.2
C	compression ratio in discussion of tamped criticality (Sects. 2.3 and 2.5); number of bare-threshold critical masses in bomb core (Sect. 2.4); constant of integration in electromagnetic separation of isotopes analysis (Sect. 3.4); constant in analysis of barrier diffusion (Sect. 3.5); solar constant (Sect. 5.2)
c	speed of light
D	diffusion coefficient (2.20); constant of integration in electromagnetic separation of isotopes analysis (Sect. 3.4)
d	product-nuclei separation in simulation of fission barrier (Sect. 1.11); characteristic length in criticality calculations, sometimes with subscripts *core* and *tamp* for core and tamper; fundamental definition in (2.25)
d_{AU}	distance in Astronomical Units (Sect. 5.2)
d_O	characteristic length in criticality calculations as d above but for case of bare threshold criticality ($\alpha = 0$)
d_{pc}	distance in parsecs (Sect. 5.2)
E	energy; often meaning a mc^2 rest energy; may be accompanied with a subscript to designate a particular particle; constant of integration in electromagnetic separation of isotopes analysis (Sect. 3.4)
$E_{barrier}$	fission barrier energy
E_f	energy liberated per fission reaction, typically ~180 MeV
$E_m\ mc^2$	rest energy
E_γ	pre-collision gamma-ray energy in analysis of neutron discovery (Sect. 1.4)
$E_\gamma{}^*$	post-collision gamma-ray energy in analysis of neutron discovery (Sect. 1.4)
e	electron charge
F	parameter in analysis of fission barrier (1.89); logarithm of number of fissions by which time the nuclear explosion within a core has become established (Sect. 4.3); fractional abundance of ^{235}U in reactor criticality calculation (Sect. 3.1); isotope fractional abundance in boron-contamination analysis (Sect. 4.1); spontaneous fission rate (4.4) in Sect. 4.2; isotope fractional abundance in reactor fuel (Sect. 5.3)
f	mass ratio of products in fission analysis (1.45) in Sect. 1.7
$f_{\text{linear}}(x)$	function in analysis of fission barrier (1.87) in Sect. 1.10
$f_{\text{smooth}}(x)$	function in analysis of fission barrier (1.89) and (1.90) in Sect. 1.10
$f_S(\gamma)$	surface-area function in simulation of fission barrier (1.107)
$f_V(\gamma)$	volume function in simulation of fission barrier (1.101)
g	parameter in predetonation analysis (4.10)

h	Planck's constant; final height of distorted nucleus in simulation of fission barrier (1.109); parameter in Newton's Law of Cooling (Sect. 5.1)
i	electrical current
J_{km}	m'th zero of Bessel function of order k (Sect. 2.9)
K	kinetic energy; often accompanied with a subscript to designate a particular particle normalization; constant in analysis of secondary-neutron energy spectrum (1.67)
K_x, K_y	constants of integration in electromagnetic separation of isotopes analysis (Sect. 3.4)
k	reproduction constant in reactor criticality, sometimes with subscript *slow* or *fast* (Sect. 3.1)
k_B	Boltzmann's constant
k_z	separation constant in analysis of cylindrical criticality (2.152) and (2.163)
k_ϕ	separation constant in analysis of cylindrical criticality (2.155) and (2.169)
L	length of cylindrical bomb core (Sect. 2.9); side length of magnet coil (Sect. 3.4); stellar luminosity (Sect. 5.2)
M_{core}	mass of bomb core
m	mass; astronomical apparent magnitude (Sect. 5.2) with subscripts A and B
N	sometimes used as number of neutrons; number of cells in simulation of fission barrier (Sect. 1.11); neutron number density in criticality and diffusion calculations, sometimes with super/sub-scripts (*core, tamp*) for core and tamper and (r, t) for radial and time-dependences (Chap. 2); number of scatterings in neutron thermalization (Sect. 3.2); equivalent number of Suns of *Trinity* brightness (Sect. 5.2); number of atoms in fuel for reactor simulation (Sect. 5.3)
N_A	Avogadro's number
N_{esc}	number of neutrons that penetrate through a sample (Sect. 2.1); number of neutrons in "escape cylinder" in diffusion and diffusion-equation calculations (Sects. 3.5, 6.7)
N_O	number of incident neutrons bombarding a sample (Sects. 2.1, 2.2, 2.4 and 2.5); initial number density of neutrons in center of bomb core (2.22)
n	nuclear number density (2.1)
n_H	nuclear number density of heavy-element fissile material in impurity calculations (Sect. 4.4)
n_L	nuclear number density of light-element impurity (Sect. 4.4)
P	pressure, usually within bomb core as (2.87); sometimes used for probability
$P(d\Omega)$	probability of particle moving in direction of solid angle $d\Omega$; (3.49)
$<P_{sph}>$	probability of neutron escape from a sphere; defined in (4.9) (Sects. 4.2, 6.6)

P_O	initial pressure at center of bomb core (2.87)
$P_{direct\ escape}$	probability of a single neutron passing through a sample to escape; (2.7) in Sect. 2.1
P_e	reactor electrical power output (Sect. 3.3)
P_{escape}	probability of neutron passing through layer of material (Sect. 2.1); neutron-escape probability in predetonation analysis (Sects. 4.2, 4.3)
$P_n(\cos\theta)$	Legendre polynomial of order n (Sect. 6.5)
P_{react}	probability that incident particle will initiate reaction (Sect. 2.1)
$P_{no\ predet}$	non-predetonation probability (Sects. 4.2, 4.3)
P_t	reactor thermal power output (Sects. 3.3, 4.5)
p	momentum
Q	energy released or consumed in a reaction; fundamental definitions in (1.2) and (1.10)
q	electric charge
R	radius of product nucleus in simulation of fission barrier (Sect. 1.11); ion orbital radius in electromagnetic isotope separation (3.40) in Sect. 3.4; rate of fissions in reactor (Sects. 3.3 and 5.3); carbon-to-uranium number ratio in analysis of neutron reproduction in graphite (Sect. 4.1)
R_α	rate of alpha-decay (Sect. 4.4)
R_C	general critical radius (2.29)
R_{core}	general core radius (Chap. 2)
R_n	rate of neutron production from (α, n) reactions (Sect. 4.4)
R_O	radius of initial nucleus in fission analysis (Sects. 1.7, 1.11, 6.5); rate of neutron bombardment per square meter (Sect. 2.1); bare threshold critical radius (Sect. 2.2)
R_{thresh}	threshold critical radius for a tamped core (Sects. 2.3, 2.4)
R_1, R_2	radii of product nuclei in fission analysis (Sect. 1.7); inner and outer core radii in analysis of composite cores (Sect. 2.4)
r	general symbol for radius, often of a bomb core as a function of time
S	as a subscript, generally meaning surface area surface area of membrane hole in gaseous diffusion analysis (Sect. 3.5); stopping power of material against alpha-particles (4.28) in Sect. 4.4
SF	spontaneous fission
S_{max}	maximum number of scatterings in predetonation analysis (Sect. 4.2)
T	absolute (Kelvin) temperature
$t_{1/2}$	half-life
t_{crit}	time for bomb core to reach second criticality (2.94)
t_{fiss}	time required to fission all material in bomb core (2.84). Also sometimes written $t_{fission}$
t_{fizzle}	time by which the nuclear reaction in a core can be considered to have begun if initiated at the moment of first criticality (Sect. 4.3)
t_{init}	time at which chain reaction in core is initiated after first criticality (Sect. 4.3)
t_O	time to assemble bomb core following first criticality (Sect. 4.3)
U	energy density in bomb efficiency analysis (Sect. 2.5)

U_C	Coulomb energy in fission analysis (Sects. 1.7, 1.11, 6.5)
U_S	surface energy in fission analysis (Sects. 1.7, 1.11, 6.5)
U_E, U_{total}	total electrostatic energy $U_S + U_C$ in fission analysis
u	atomic mass unit
V	volume; usually of bomb core
V_{acc}	accelerating voltage in electromagnetic separation of isotopes (Sect. 3.4)
v	speed
v_{neut}	neutron speed
v_{mp}	most probable speed (3.9)
x	fissility parameter (1.86) penetration distance in neutron escape probability (Sect. 2.1); dimensionless length parameter in criticality calculations (2.26); reduced path-length neutron travel distance in spherical escape probability calculation (6.74) in Sect. 6.6
x_1, x_2, x_3	Parameters in alysis of composite-core criticality (2.72)
Y	bomb energy yield (2.95)
y	yield of nuclear reaction, usually applied to (α, n) reactions (Sect. 4.4)
Z	atomic (proton) number

Greek letters

α alpha particle
parameter in analysis of neutron discovery (1.23);
parameter in analysis of spontaneous-fission limit (1.50);
parameter in analysis of secondary-neutron energy spectrum (1.64);
neutron-density exponential growth parameter introduced in (2.22);
separation constant in analysis of electromagnetic separation of isotopes (3.29) in Sect. 3.4;
parameter in analysis of two-body to two-body collision; in Sect. 6.3;
parameter in analysis of two-body to massive particle plus gamma-ray collision in Sect. 6.4

α_O design value of neutron-density exponential growth parameter (Sect. 4.3)
α_0, α_2 surface-distortion parameters in Bohr-Wheeler analysis of fission (Sect. 6.5)
β parameter in analysis of spontaneous-fission limit (1.55);
angle in simulation of fission barrier (Sect. 1.11);
parameter in Peierls' analysis of criticality (2.123);
parameter in analysis of two-body to two-body collision in Sect. 6.3

γ parameter in analysis of spontaneous-fission limit (1.56);
angle in simulation of fission barrier (Sect. 1.11);
gas/radiation pressure constant (Sects. 2.5 and 2.6);
parameter in analysis of two-body to two-body collision (Sect. 6.3)
Δ mass excess in MeV (1.10)
ΔE fission barrier

Δr bare-core expansion distance to second criticality (2.91)

δ parameter in analysis of neutron discovery (1.25);
neutron-density exponential growth parameter for tamper material (Sect. 2.3);
parameter in analysis of composite-core criticality (Sect. 2.4);
parameter in analysis of two-body to massive particle plus gamma-ray collision (Sect. 6.4)

ε parameter in analysis of neutron discovery (1.24);
parameter in analysis of energy distribution of fission-liberated neutrons (1.71);
parameter in criticality calculations (2.31);
parameter in analysis of two-body to massive particle plus gamma-ray collision(Sect. 6.4)

ε_O permittivity of free space

η numerical factor in simulation of fission barrier related to final height of nucleus (1.109)

θ polar angle in simulation of fission barrier (Sects. 1.11, 6.5)

κ parameter in analysis of cylindrical criticality (2.158)

λ mean free path; super/subscript indicates process type (*fission, capture, scattering...*) and *core* or *tamper* if necessary
parameter in analysis of composite-core criticality; Sect. 2.4

μ mass excess (Sect. 1.1)
average number of spontaneous fissions in predetonation analysis (Sects. 4.2, 4.3)

μ_O permeability of free space (Sect. 3.4)

ν number of secondary neutrons liberated per fission

ξ parameter in Peierls' analysis of criticality (2.118)

Φ neutron flux(Sects. 3.3 and 4.5)

π_k probability of neutron escape following k scatterings in predetonation analysis (Sect. 4.2)

ρ mass density; charge density; occasionally used as a number density;
reduced radial neutron travel distance in spherical escape probability calculation (6.73) in Sect. 6.6

Σ surface area of sample under neutron bombardment (Sect. 2.1);
parameter in analysis of neutron escape from within a sphere (Sect. 6.6)

Σ_C Coulomb-energy function in simulation of fission barrier (1.115)

Σ_S surface-energy function in simulation of fission barrier (1.111)

σ reaction cross-section; subscript indicates reaction type (*fission, capture, scattering...*)

τ neutron travel time between fissions (2.21)

6.10   Appendix J: Further Reading

An online search keyed on the phrase "Manhattan Project" will typically return millions of hits. While many of the sources that turn up are interesting and well-researched, it can be difficult to sort through such a deluge of material for credible, objective information on the history, science, and personalities associated with the Project, particularly for readers new to the topic. This Appendix offers a very selective annotated bibliography of Manhattan Project books, articles, journal papers, and websites. Some of the references listed here appear in various chapters in this book but are copied here for sake of completeness. More complete annotated bibliographies can be found in three Resource Letters published by the author in *American Journal of Physics*, **73**(9), 805-811 (2005); **79**(2), 151-163 (2011); and **84**(10) 734-745 (2016).

The most important source of primary information on the Manhattan Project is the Manhattan District History (MDH). This extensive (almost 14,000 pages) multi-volume document was previously available only on microfilm, but is now available online at https://www.osti.gov/opennet/manhattan_district.jsp. This document was prepared as an official history of the Project after the war by Gavin Hadden, an aide to General Leslie Groves, the Commanding General of the Project. The online version of the MDH also includes previously-redacted material on the highly-classified K-25 gaseous diffusion plant (Sect. 3.5).

The sources cited below are divided into four categories: general works, biographical and autobiographical works, technical works, and websites. Web addresses appear in *italics* to discriminate them from surrounding text.

6.10.1   General Works

Coster-Mullen, J.: Atom Bombs: The Top Secret Inside Story of Little Boy and Fat Man (2016). This remarkable self-published work contains a trove of drawings, photographs, reproductions of documents, mission logs, reports, details of *Little Boy* and *Fat Man*, and details of the Hiroshima and Nagasaki bombing missions. Constantly being updated; available from online booksellers.

Fermi, R., Samra, E.: Picturing the Bomb: Photographs from the Secret World of the Manhattan Project. Harry N. Abrams, New York (1995). Beautifully reproduced and instructively captioned photographs of sites and artifacts associated with the MP. The first author is Enrico Fermi's granddaughter.

Gibson, T.M., Michnovicz, J.: Los Alamos 1944–1947 (Arcadia, Charleston, SC, 2005). In October, 1944, Private John Michnovicz arrived as Los Alamos to serve as a photographer. Over the following three years he took thousands of photos of residents, famous and anonymous alike, at work and play. This book, part of the Images of America series (see Westcott below), was prepared by Michnovicz's daughter and son and reproduces several dozen shots of scenery, social and recreational events, pets, and formal portraits of laboratory staff, all informatively captioned.

Gosling, F.G.: The Manhattan Project: Making the Atomic Bomb. U.S. Department of Energy (1999). A brief but very readable and well-illustrated summary. Available free from the DoE at https://www.energy.gov/management/downloads/gos ling-manhattan-project-making-atomic-bomb.

Hersey, J.: Hiroshima. Knopf, New York (1985). Originally published in 1946, this compelling work of firsthand accounts of Hiroshima survivors is a must-read for students of the MP. The edition cited here includes an additional chapter written 40 years later which brings the survivors' stories up-to-date.

Hewlett, R.G., Anderson, O.E.: A History of the United States Atomic Energy Commission, Vol. 1: The New World, 1939/1946. Pennsylvania State University, University Park, PA (1962). Detailed history of the Manhattan Project prepared under the auspices of the Historical Advisory Committee of the U. S. Atomic Energy Commission. Fully referenced to Manhattan Engineer District documents.

Jones, V.C.: United States Army in World War II. Special Studies. Manhattan: The Army and the Atomic Bomb. Center of Military History, United States Army, Washington, DC (1985). A comprehensive history of Army involvement in the Manhattan Project, fully referenced to Manhattan Engineer District documents. Unfortunately, this work no longer appears to be available through the Government Printing Office, but many libraries have copies.

Laurence, W.L.: Dawn Over Zero: The Story of the Atomic Bomb. Knopf, New York (1946). Laurence was a *New York Times* science reporter who was allowed to visit Los Alamos during the Project, witnessed the *Trinity* test, and rode aboard the instrumentation plane *The Great Artiste* during the Nagasaki bombing mission. This work was the one of the first serious popular accounts of the Project.

Manhattan Engineer District: The Atomic Bombings of Hiroshima and Nagasaki (1946). This report, dated June 29, 1946, summarizes the results of investigative teams dispatched to Hiroshima and Nagasaki in the weeks following the end of the war; reprinted copies are available from various online vendors; see also http://www. atomicarchive.com/Docs/MED. Surveys the nature of the cities before the bombings, casualty rates, percentages of structures destroyed, and blast, thermal, and radiation effects on both structures and people.

Reed, B.C.: The History and Science of the Manhattan Project, 2nd ed. Springer, Berlin (2019). This book, which can be regarded as a companion volume to the present work, covers the technical and administrative history of the Project at a semi-popular level from its beginnings through the bombings of Hiroshima and Nagasaki, and offers a brief survey of postwar weapons developments and deployments. This edition includes a chapter on the wartime German nuclear program. Highly recommended.

Reed, B.C.: Manhattan Project: the Story of the Century. Springer, Berlin (2020). An excellent popular treatment of the Manhattan Project.

Rhodes, R.: The Making of the Atomic Bomb. Simon and Schuster, New York (1986). This work is an outstanding overall survey of the context, personalities, and history of the Manhattan Project. Contains an extensive bibliography. A follow-on book, Dark Sun: The Making of the Hydrogen Bomb (Simon and Schuster, New York, 1995) details the development of the hydrogen bomb and is particularly interesting for its description of Soviet espionage in the United States during the Manhattan Project and afterward.

Russ, H.W.: Project Alberta: The Preparation of Atomic Bombs for use in World War II. Exceptional Books, Los Alamos (1990). Project Alberta was that part of the Manhattan Project responsible for assembly and delivery of the working weapons. Harlow Russ was an engineer at Los Alamos who was involved with the design of the *Fat Man* implosion mechanism and who shipped out to Tinian Island in June, 1945. This well-illustrated volume gives a first-person account of the final preparations of the weapons.

Smyth, H.D.: Atomic Energy for Military Purposes: The Official Report on the Development of the Atomic Bomb under the Auspices of the United States Government, 1940–1945. Princeton University Press, Princeton, NJ, (1948). This work was the first official report on the Manhattan Project. The edition cited here includes various appendices not included in the original 1945 edition. Various editions are readily available online.

Stoff, M.B., Fanton, J.F., Williams, R.H.: The Manhattan Project: A Documentary Introduction to the Atomic Age. McGraw Hill, New York (1991). This book includes reproductions of a number of official documents and memoranda concerning the Project. It is now somewhat dated because so much material is available online, but it is still worth perusing.

Stuewer, R.: The Age of Innocence: Nuclear Physics between the First and Second World Wars. Oxford University Press, Oxford, (2018). A detailed history of the remarkable experimental, theoretical, and instrumental developments in nuclear physics between the World Wars. Also contains capsule biographies of many of the leading personalities, covering both their scientific accomplishments and personal lives. Extensive bibliography and references; well-illustrated.

Toomey, E.: Images of America: The Manhattan Project at Hanford Site. Arcadia, Charleston, SC (2015). This book is another volume in the Images of America series, featuring photos of life and work at Hanford which document the scale of construction and operations and aspects of day-to-day life in trying conditions. Many of the photos were taken by Robley Johnson, who supervised DuPont's photography crew at the site.

United States Department of Energy: The First Reactor (1982). This publication presents a brief, well-illustrated account of Enrico Fermi's CP-1 reactor. Available online at http://www.osti.gov/accomplishments/documents/fullText/ACC0044.pdf.

Westcott, E.: Images of America: Oak Ridge. (Arcadia, Charleston, SC, 2005). Ed Westcott was an official Manhattan Project photographer who enjoyed complete access to all aspects of life and work at Oak Ridge. This book of his photographs illustrates not only the frantic pace of construction and work at the Clinton Engineer Works, but also scenes of family life, day-to-day activities, and recreation in what became a city of 75,000 people by the end of the war. A website with Westcott's photos can be found at *photosofedwestcott.tumblr.com*

Williams, H.: Made in Hanford: The Bomb that Changed the World. Washington State University Press, Pullman, WA (2011). This book describes how the area around

Hanford, Washington, came to be chosen as the site of the Manhattan Project's plutonium production reactors. Williams describes how the town was transformed by the project and the engineering and operation of the reactors.

6.10.2   Biographical and Autobiographical Works

Bernstein, J.: Hitler's Uranium Club: The Secret Recordings at Farm Hall. American Institute of Physics, New York (1996). At the end of the war, a number of leading German nuclear physicists including Werner Heisenberg were interned for six months at Farm Hall, an English country estate, and their conversations secretly recorded. Bernstein analyses the transcripts; particularly interesting is the captives' response to the news of the bombing of Hiroshima.

Bernstein, J.: Oppenheimer: Portrait of an Enigma. Ivan R. Dee, Inc., Chicago (2004). Engaging brief biography of Oppenheimer; Bernstein knew him personally.

Bird, K., Sherwin, M.J.: American Prometheus: The Triumph and Tragedy of J. Robert Oppenheimer. Knopf, New York (2005). This book is the definitive biography of Oppenheimer. The authors particularly examine his upbringing, ethical outlook, and postwar political activities. Descriptions of physics is muddled in some places.

Cassidy, D.C.: Uncertainty: The Life and Science of Werner Heisenberg. W.H. Freeman, New York (1993). Published prior to release of the Farm Hall transcripts [see Bernstein (1996) above], this work continues to be the major scholarly biography of Heisenberg. In 2009, Cassidy published a successor volume, Beyond Uncertainty: Heisenberg, Quantum Physics, and the Bomb (Bellevue Literary Press, New York) which incorporates much new material.

Cassidy, D.C.: J. Robert Oppenheimer and the American Century. Pi Press, New York (2005). A full scholarly biography of Oppenheimer. Cassidy devotes fairly little space to the well-trodden ground of Oppenheimer's Los Alamos years, but gives a much more complete picture of his life and scientific work than many sources. Includes lists of Oppenheimer's publications and students.

Fermi, L.: Atoms in the Family: My Life with Enrico Fermi. University of Chicago Press, Chicago (1954). Fermi's life and work as related by his wife, Laura. Chapters 18–23 deal with the first chain-reacting pile and the Fermis' time at Los Alamos. See also Segrè and Schwartz below.

Frisch, O. What Little I Remember. Cambridge University Press, Cambridge (1979). Frisch helped to interpret fission, is generally credited with being the first experimenter to deliberately produce fission, collaborated with Rudolf Peierls to produce the first estimate of the critical mass of ^{235}U, and worked as Los Alamos during the war.

Goodchild, P.: Robert Oppenheimer: Shatterer of Worlds. BBC, London, (1980). A well-illustrated treatment of the Manhattan Project and Oppenheimer's life; this book was the companion volume to a BBC television series.

Groves, L.R.: Now It Can Be Told: The Story of the Manhattan Project. Harper & Row, New York (1962). Now somewhat dated, but still valuable; the view from one who was there.

Howes, R.C., Herzenberg, C.C.: Their Day in the Sun: Women of the Manhattan Project. Temple University Press, Philadelphia (1999). Examines the lives and work of female physicists, chemists, biologists, technicians and others on the Project.

Kiernan, D.: The Girls of Atomic City: The Untold Story of the Women Who Helped Win World War II. Touchstone, New York (2013). This engaging volume tells the stories of a number of women who worked at Oak Ridge in capacities such as secretaries, calutron operators, chemists, janitors, gaseous-diffusion plant leak detectors, and statisticians. Kiernan puts a very human perspective on the lives and tribulations of the thousands of employees that kept the enormous enrichment facilities of the Clinton Engineer Works operating.

Norris, R.S.: Racing for the Bomb: General Leslie R. Groves, The Manhattan Project's Indispensable Man. Steerforth Press, South Royalton, VT (2002). Definitive account of the life and work of General Groves.

Pais, A., Crease, R.P.: J. Robert Oppenheimer: A Life. Oxford, New York (2006). Pais knew Oppenheimer from 1946 until the latter's death in 1967 and had completed about three-quarters of this work before his own passing in August, 2000; the book was completed by his widow and Robert Crease. The emphasis is not so much on Oppenheimer's Los Alamos years, but rather on his contributions to the growth of American theoretical physics, his postwar directorship of the Institute for Advanced Study, as a leader of conferences, and his service on numerous government committees.

Peierls, R.: Bird of Passage: Recollections of A Physicist. Princeton University Press, Princeton (1985). Memoirs of the other half of the Frisch-Peierls team; written with humor and warmth.

Segrè, E.: Enrico Fermi: Physicist. University of Chicago Press, Chicago (1970). Very readable biography by one of Fermi's closest collaborators.

Segrè, G., Hoerlin, B.: The Pope of Physics: Enrico Fermi and the Birth of the Atomic Age. Henry Holt, New York (2016). This excellent biography of Fermi is written by Gino Segrè and his wife Bettina Hoerlin; Gino is the nephew of Fermi's student Emilio Segrè.

Schwartz, D.N.: The Last Man Who Knew Everything: The Life and Times of Enrico Fermi, Father of the Nuclear Age. Basic Books, New York (2017). An excellent biography of Fermi.

Sime, R.L.: Lise Meitner: A Life in Physics. University of California Press, Berkeley (1996). By far the best account of the life and work of Lise Meitner, who along with Robert Frisch interpreted the meaning of fission. Contains useful appendices explaining radioactive decay sequences.

6.10.3 Technical Works

Amaldi, E.: From the Discovery of the Neutron to the Discovery of Nuclear Fission. Physics Reports **111** (1–4), 1–331 (1984). Masterful account of the development of nuclear physics during the 1930's; contains over 900 references. Amaldi's career began as a student of Fermi.

Bernstein, J.: Plutonium: A History of the World's Most Dangerous Element. Joseph Henry Press, Washington (2007). Reviews the history of the discovery and manufacture of plutonium, its bizarre chemical properties, and the important but often-overlooked contributions of metallurgists at Los Alamos.

Bernstein, J.: Nuclear Weapons: What You Need to Know. Cambridge University Press, Cambridge, UK (2008). This companion volume to the above entry summarizes the development of nuclear weapons from the discoveries of Thomson and Rutherford through the first North Korean test of 2006. Full of interesting personal anecdotes and sidebar stories.

Bernstein, J.: A Memorandum that Changed the World. Am. J. Phys. **79**(5), 440–446 (2011). Bernstein analyzes the Frisch-Peierls memorandum of early 1940 which first brought the possibility of fission weapons to the attention of British government officials. Frisch and Peierls drastically underestimated the critical mass of ^{235}U, but their analysis was fundamentally correct.

Brode, H.L.: Review of Nuclear Weapons Effects. Ann. Rev. Nuc. Sci. **18**, 153–202 (1968). This article gives an advanced technical account of the fireball, shock, thermal, radiation, electromagnetic pulse, and fallout effects of nuclear explosions.

Broyles, A.A.: Nuclear Explosions. Am. J. Phys. **50**(7), 586–594 (1982). An undergraduate-level account of the effects of nuclear explosions.

Fermi, E.: Experimental Production of a Divergent Chain Reaction. Am. J. Phys. **20**(9), 536–558 (1952). Description of Fermi's first critical pile, published on the tenth anniversary of that achievement.

Garwin, R.L., Charpak, G.: Megawatts and Megatons: A Turning Point in the Nuclear Age? Knopf, New York (2001). Excellent treatment of nuclear power, nuclear weapons, radiation effects, waste disposal, and associated environmental and political issues.

Glasstone, S., Dolan, P. J.: The Effects of Nuclear Weapons, 3rd edition. United States Department of Defense and Energy Research and Development Administration, Washington (1977). This sobering volume summarizes technical analyses of the shock, blast, thermal, and radiation effects of nuclear explosions on structures and people. Available as an electronic book on CD from http://www.amazon.com/The-Effects-Nuclear-Weapons-Authoritative/dp/1422050645.

Guerra, F., Leone, M., Robotti, N.: The Discovery of Artificial Radioactivity. Phys. Perspect. **14**(1), 33–58 (2012). This paper examines the circumstances of the discovery of artificial radioactivity by the Joliot-Curies' in early 1934. Can be read as a complement to Amaldi (1984) above.

Hawkins, D.: Project Y, the Los Alamos Story. Tomash, Los Angeles, (1983). Originally published as Los Alamos report LAMS-2532, Manhattan District History,

Project Y, The Los Alamos Project. Now unfortunately out of print, this book gives a detailed technical and administrative history of Los Alamos from its inception through December, 1946. The original Los Alamos report on which the book is based is available from the Federation of American Scientists at http://www.fas.org/sgp/othergov/doe/lanl/docs1/00103803.pdf.

Hoddeson, L., Henriksen, P.W., Meade, R.A., Westfall, C.: Critical Assembly: A Technical History of Los Alamos during the Oppenheimer Years, 1943–1945. Cambridge University Press, Cambridge (1993). An authoritative technical history of Los Alamos during the war years.

L'Annunziata, M.: Radioactivity: Introduction and History. Elsevier Science, Amsterdam (2007). An excellent survey of the history of discoveries in radiation and radioactive decay, with emphasis on original works; enlivened with biographical accounts.

Reed, B.C. Arthur Compton's 1941 report on explosive fission of U-235: A look at the physics. American Journal of Physics **75**(12), 1065–1072 (2007). Technical material pertinent to Sect. 2.2 of the present book.

Reed, B.C. A Brief Primer on Tampered Fission-Bomb Cores. American Journal of Physics **77**(8), 730–733 (2009). This companion paper to the entry immediately above explores the physics of tamped bomb cores as discussed in Sect. 2.3 of the present book.

Reed, B.C. Student-level numerical simulation of conditions inside an exploding fission-bomb core. Natural Science **2**(3), 139–144 (2010). Material relevant to Sect. 2.6 of the present book.

Reed, B.C. Predetonation probability of a fission-bomb core. American Journal of Physics. **78**(8), 804–808 (2010). Material relevant to Sect. 4.2 of the present book.

Reed, B.C. A desktop-computer simulation for exploring the fission barrier." Natural Science **3**(4) 323–327 (2011). Material relevant to Sect. 1.11 of the present book.

Reed, B.C. Fission fizzles: Estimating the yield of a predetonated nuclear weapon. American Journal of Physics, **79**(7), 769–773 (2011). This companion paper to the predetonation probability paper two entries above extends the analysis to explore the physics of predetonation yield as in Sect. 4.3 of the present book.

Reed, B.C. Liquid Thermal Diffusion during the Manhattan Project. Physics in Perspective **13**(2), 161–188 (2011). The S-50 liquid thermal diffusion uranium-enrichment project has tended to be overlooked in comparison with its gargantuan K-25 and Y-12 counterparts. This paper explores the history of this project at a semi-popular level.

Reed. B.C.: From Treasury Vault to the Manhattan Project. American Scientist **99**(1), 40–47 (2011). Over 14,000 t of silver were borrowed from the U.S. Treasury to make magnet coils for the calutron electromagnetic isotope separators at Oak Ridge. This paper relates the history of this little-known part of the Project.

Reed, B.C.: A Physicist's Guide to *The Los Alamos Primer*, Phys. Scr. **91**(11) 113002 (30 pp) (2016). Erratum: Phys. Scr. **91**(12) 129601 (1p) (2016). This paper offers a detailed analysis of the physics of Robert Serber's *Los Alamos Primer* (see immediately below).

Serber, R.: The Los Alamos Primer: The First Lectures on How To Build An Atomic Bomb. University of California Press, Berkeley (1992). Copies of lectures given by Serber to scientists arriving at Los Alamos in April 1943 are reproduced and supplemented by extensive annotations. Includes a copy of the March 1940 Frisch-Peierls memoranda that can be said to have started the Project.

6.10.4 Websites

Readers are cautioned that websites and addresses can change.

ALSOS digital library for nuclear issues: The U. S. Nuclear Weapons Complex, https://www.google.com/maps/d/viewer?mid=16D-GF2of9UXppSRknAN_ApFp HBg&ll=27.322007519566913%2C-85.61426514999994&z=2. This interactive map contains the locations of dozens of offices, mines, mills, plants, laboratories, and test sites that comprised the U.S. nuclear complex from World War II to 2016. Sections are organized to follow the progression of nuclear materials from mining to weapons testing. A unique color code identifies the locations in each section.

Carey Sublette maintains an extensive site on nuclear weapons at http://nuclea rweaponarchive.org/Nwfaq/Nfaq8.html.

Historian of science Alex Wellerstein maintains a website on nuclear secrecy, *nuclearsecrecy.com/blog*, which contains a number of interesting documents, photos, and resources for teachers and students. Of particular note is his "Nukemap" program, which allows users to get an idea of the effects of detonating a nuclear weapon of specified yield at a given location; *nuclearsecrecy.com/nukemap*

Secret City: The Oak Ridge Story. The War Years. HP Video (2005). This DVD relates the story of the development of the Clinton Engineer Works at Oak Ridge, particularly the Y-12 and K-25 plants and the X-10 pilot-scale graphite reactor. Rare color footage of construction activities is interspersed with black-and-white images of community and social life and interviews with people who worked there. Also, a 9-minute video of some of the earliest known footage of construction at Oak Ridge is available at http://www.youtube.com/watch?v=HWnm4N8Wnmk& feature=channel_video_title.

The Los Alamos National Laboratory's history website can be found at http:// www.lanl.gov/history.

The *Bulletin of the Atomic Scientists* is a good source of up-to-date information on weapons deployments, treaties concerning nuclear weapons, and nuclear issues in general. www.thebulletin.org.

The National Science Foundation Digital Library on the Atomic Bomb at www. atomicarchive.com contains material on the history and science of the atomic bomb, and includes links to the full text of the Smyth report and the declassified version of Bainbridge's report on the Trinity test.

The office for history of science and technology at the University of California at Berkeley has created a website exploring Robert Oppenheimer's life. http://ohst.ber keley.edu.

The homepage of the National Museum of Nuclear Science and History can be found at http://www.nuclearmuseum.org.

The U.S. Department of Energy Office of History and Heritage Resources maintains an Interactive History website on the Manhattan Project at http://energy.gov/search/site/Manhattan%20Project?gid=49.

A project at Lawrence Livermore National Laboratory is dedicated to digitizing films of the laboratory's atmospheric nuclear weapons tests. Freely available at https://www.youtube.com/playlist?list=PLvGO_dWo8VfcmG166wKRy5z-GlJ_OQND5

The Nevada Site Office of the National Nuclear Security Administration offers an online collection of films of nuclear tests conducted between 1945 and 1962 https://www.nnss.gov/pages/resources/library/Media.html.

The Harry S. Truman Library and Museum makes available online a collection of documents, diary entries, letters and press releases relevant to Truman's involvement with atomic bombs: http://www.trumanlibrary.org/whistlestop/study_collections/bomb/large.

The Atomic Heritage Foundation is a non-profit organization dedicated to the preservation and interpretation of the Manhattan Project, and works with the Department of Energy and former Manhattan Project communities to preserve historic resources and other aspects of Project history. www.atomicheritage.org. The Foundation's website "Voices of the Manhattan Project" features oral histories of Manhattan Project veterans; http://manhattanprojectvoices.org.

The Federation of American Scientists maintains a website with links to copies of hundreds of Los Alamos Technical Reports and Publications. http://www.fas.org/sgp/othergov/doe/lanl/index1.html.

The B-Reactor Museum Association is a volunteer group that works with local, state and federal authorities to preserve the Hanford B reactor and turn it into a publicly-accessible museum. http://www.b-reactor.org.

6.11   Appendix K: Useful Constants and Rest Masses

Values are adopted from the National Institute of Standards and Technology; https://physics.nist.gov/cuu/pdf/wall_2018.pdf.

Quantity	Symbol	Value	Unit
Speed of light	c	2.99792458×10^8	$m\ s^{-1}$
Electron charge	e	$1.602176634 \times 10^{-19}$	C
Planck's constant	h	$6.62607015 \times 10^{-34}$	Js
Permittivity constant	ε_O	$8.85418782 \times 10^{-12}$	$C^2\ (Jm)^{-1}$
Avogadro's number	N_A	$6.02214076 \times 10^{23}$	mol^{-1}
Atomic mass unit	u	$1.660539067 \times 10^{-27}$	kg

(continued)

(continued)

Quantity	Symbol	Value	Unit
Boltzmann constant	k_B	1.380649×10^{-23}	$J\,K^{-1}$
Kiloton TNT	kt	4.2×10^{12}	J
Fission 1 kg ^{235}U		7.1×10^{13}	J
Fission 1 kg ^{235}U		~17	kt
Curie	Ci	3.7×10^{10}	$decay\ s^{-1}$
Calorie	cal	4.186	J
Barn	bn	1.0×10^{-28}	m^2

Rest masses

amu $= 1.6605390666 \times 10^{-27}$ kg

	10^{-27} kg	amu	MeV
Proton	1.67262192	1.00727646662	938.272088
Neutron	1.67492750	1.00866491595	939.565420
Electron	$9.109383702 \times 10^{-4}$	$5.485799091 \times 10^{-4}$	0.510998950
Alpha	6.64465734	4.0015061791	3727.37941

References

Bernstein, J., Pollock, F.: The calculation of the electrostatic energy in the liquid drop model of nuclear fission–A pedagogical note. Physica **96A**, 136–140 (1979)

Bernstein, J.: Plutonium: A History of the World's Most Dangerous Element. Joseph Henry Press, Washington (2007)

Bernstein, J.: Nuclear Weapons: What You Need to Know. Cambridge University Press, New York (2008)

Bohr, N., Wheeler, J.A.: The mechanism of nuclear fission. Phys. Rev. **56**, 426–450 (1939)

Cottingham, W.N., Greenwood, D.A.: An Introduction to Nuclear Physics, 2nd edn, p. 84. Cambridge University Press, Cambridge (2001)

Croft, S.: The distribution of emergent flight path lengths from homogeneous spheres. Nucl. Instrum. Meth. Phys. **A288**, 589–592 (1990)

Feynman, R.P., Leighton, R.B., Sands, M.: The Feynman Lectures on Physics. Addison-Wesley, New York (1964)

Liverhant, S.E.: Elementary Introduction to Nuclear Reactor Physics. Wiley, New York (1960)

Plesset, M.S.: On the classical model of nuclear fission. Am. J. Phys. **9**, 1–10 (1941)

Present, R.D., Knipp, J.K.: On the dynamics of complex fission. Phys. Rev. **57**, 751 (1940a)

Present, R.D., Knipp, J.K.: On the dynamics of complex fission. Phys. Rev. **57**, 1188–1189 (1940b)

Reed, B.C.: The Bohr-Wheeler spontaneous fission limit: an undergraduate-level derivation. Eur. J. Phys. **30**, 763–770 (2009)

Reed, B.C.: The History and Science of the Manhattan Project, 2nd edn. Springer, Berlin (2019)

Serber, R.: The Los Alamos Primer: The First Lectures on How To Build An Atomic Bomb. University of California Press, Berkeley (1992)

Index

© The Editor(s) (if applicable) and The Author(s), under exclusive license to
Springer Nature Switzerland AG 2021
B. C. Reed, *The Physics of the Manhattan Project*,
https://doi.org/10.1007/978-3-030-61373-0

Printed in the United States
by Baker & Taylor Publisher Services